Elements

NITROGEN AND PHOSPHORUS

How to use this book

This book has been carefully developed to help you understand the chemistry of the elements. In it you will find a systematic and comprehensive coverage of the basic qualities of each element. Each two-page entry contains information at various levels of technical content and language, along with definitions of useful technical terms, as shown in the thumbnail diagram to the right. There is a comprehensive glossary of technical terms at the back of the book, along with an extensive index, key facts, an explanation of the Periodic Table, and a description of how to interpret chemical equations.

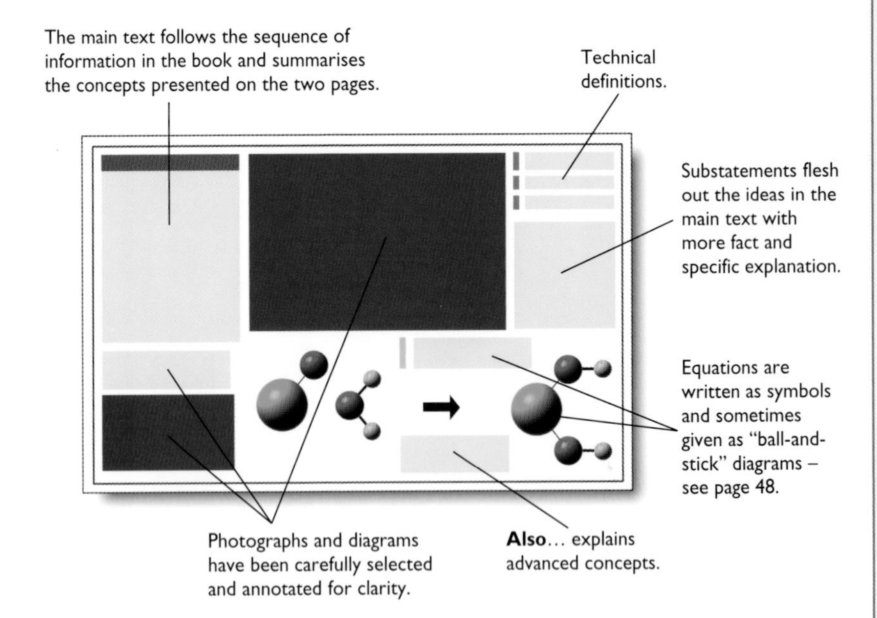

An Atlantic Europe Publishing Book

Author
Brian Knapp, BSc, PhD
Project consultant
Keith B. Walshaw, MA, BSc, DPhil
(Head of Chemistry, Leighton Park School)
Industrial consultant
Jack Brettle, BSc, PhD (Chief Research Scientist, Pilkington plc)
Art Director
Duncan McCrae, BSc
Editor
Elizabeth Walker, BA
Special photography
Ian Gledhill
Illustrations
David Woodroffe and David Hardy
Electronic page make-up
Julie James Graphic Design
Designed and produced by
EARTHSCAPE EDITIONS
Print consultants
Landmark Production Consultants Ltd
Reproduced by
Leo Reprographics
Printed and bound by
Paramount Printing Company Ltd

First published in 1996 by
Atlantic Europe Publishing Company Limited, Greys Court Farm, Greys Court, Henley-on-Thames, Oxon, RG9 4PG, UK.

Suggested cataloguing location
Knapp, Brian
Nitrogen and Phosphorus
ISBN 1 869860 59 4
– *Elements* series
540

Acknowledgements
The publishers would like to thank the following for their kind help and advice: *ICI (UK), British Petroleum International* and *Vauxhall Motors Limited.*

Picture credits
All photographs are from the **Earthscape Editions** photolibrary except the following:
(c=centre t=top b=bottom l=left r=right)
British Petroleum Company plc 13tl; **ICI** 17tr, 17bl, 21br, 27cr; **NASA** 7tr; **SPL** 18/19, 29br; **Vauxhall Motors Limted** 29t; **Tony Waltham, Geophotos** 28bl and **ZEFA** 7cl, 8/9t, 9br, 21t, 26b.

Front cover: Nitrogen dioxide gas has a characteristic brown colour.
Title page: The head of strike-anywhere match contains a phosphorus compound that ignites with friction. This initiates a reaction of other compounds in the match head.

Contents

Introduction

An element is a substance that cannot be broken down into a simpler substance by any known means. Each of the 92 naturally occurring elements is therefore one of the fundamental materials from which everything in the Universe is made.

Nitrogen

Nitrogen, chemical symbol N, is a colourless, relatively unreactive (inert) gas without any smell. It is a kind of inert "filler" in our atmosphere, making up over three-quarters of the air we breathe.

The word nitrogen is made up from two Greek words: *nitron*, meaning saltpetre, and *genes*, meaning producing. This is because the most important historic use of nitrogen was as saltpetre, a compound called potassium nitrate, well known since ancient times as a fertiliser.

Nitrogen as a gas is relatively inert, but some nitrogen is also vital to life as part of amino acids, the building blocks of proteins. Thus, all living things are made up of compounds including nitrogen, oxygen, hydrogen and carbon. Each of these comes from the air and water around us.

Nitrogen from the air cannot be used directly by plants, so they rely on getting it from soluble nitrogen compounds instead.

Each molecule of nitrogen is made up of two nitrogen atoms linked together extremely strongly. Energy is needed to convert the nitrogen (N_2) molecules in the air into nitrogen-

containing compounds. The process of converting the nitrogen is called "fixing". Nature provides two fixing mechanisms. One occurs whenever lightning flashes and heats up the air. Another is through the efforts of the bacteria that live in the nodules of some plant roots (plants called legumes, which include peas, beans and clover). To make use of nitrogen as a fertiliser, people have to dig up rocks containing nitrogen compounds, find chemical ways to make nitrogen compounds or grow legumes in rotation with other plants.

Considerable energy is released when compounds containing nitrogen (N) atoms change and form nitrogen gas (N_2). This is why nitrogen compounds are favoured as raw materials for explosives, from gunpowder to nitroglycerine.

Phosphorus

Phosphorus, chemical symbol P, is a solid, first obtained from dried urine! It was soon discovered that phosphorus glowed in the dark and because of this property chemists named the element after the Greek word for "light-bringing".

Quite unlike nitrogen, phosphorus is very reactive and does not exist on its own in nature. On exposure to air, white phosphorus burns spontaneously. However, like nitrogen, phosphorus is known for its use in weapons, in this case because it produces intense flame. Also like nitrogen, phosphorus is essential to life, its compounds helping in the process of transferring energy within the body.

◀ Nitrogen dioxide gas being released by a reaction of fuming nitric acid and copper turnings.

Nitrogen and the atmosphere

Nitrogen is the most abundant gas in the atmosphere, making 79% by volume and 76% by mass (weight). Nitrogen molecules are bound together so tightly that nitrogen (unlike oxygen) is undisturbed by ultraviolet radiation and is found as gas molecules in the high atmosphere.

Nitrogen was one of the first gases to be formed in the early atmosphere, some four billion years ago. The most likely source was as ammonia gas coming out of volcanoes and hot springs. The ammonia was then acted on by light, breaking down to produce nitrogen and hydrogen.

Over time the other gases of the early atmosphere were partly combined with the Earth's rocks and oceans or, in the case of hydrogen, disappeared into space. As a result, the inert nitrogen was left as a greater and greater proportion of the atmosphere.

Nitrogen rarely reacts with the other gases in the atmosphere except at high temperatures during lightning flashes (and in vehicle engines). During a lightning flash, when temperatures reach 30,000°C, nitrogen and oxygen combine to make nitrogen dioxide, which, when dissolved in water, makes dilute nitric acid. This is a natural (and modest) source of acid rain. At the same time it makes nitrogen into soluble products that can be absorbed by plant roots.

Nitrogen as a source of carbon-14

In the upper atmosphere, nitrogen atoms are bombarded by cosmic rays in the form of neutrons. These change the atomic number of the nitrogen, converting nitrogen into radioactive carbon (the carbon isotope carbon-14).

Carbon-14 and oxygen atoms combine to form radioactive carbon dioxide, a gas that is absorbed by plants. In this way tiny amounts of carbon-14 are stored by plants. The residual carbon-14 in carbon-containing materials can be measured to determine the age of the material.

▼ At first the Earth had no atmosphere, but volcanic eruptions slowly produced the gases that make the atmosphere we have today, which is dominated by nitrogen.

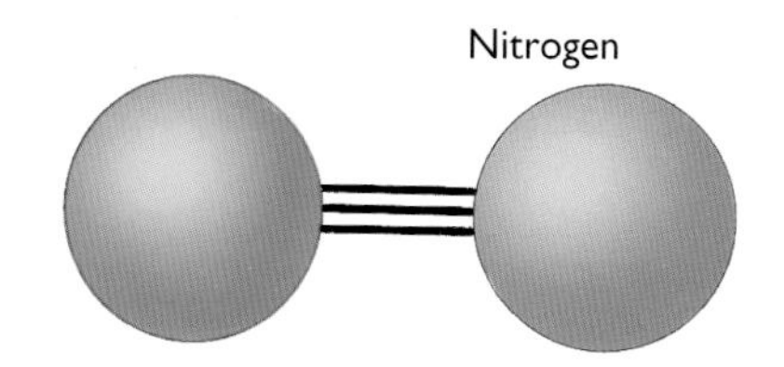

► A nitrogen molecule. The three lines joining the two atoms represent a triple bond, which shows that nitrogen has a high "bond energy".

dissolve: to break down a substance in a solution without a resultant reaction.

isotope: atoms that have the same number of protons in their nucleus, but which have different masses; for example, carbon-12 and carbon-14.

radioactive: a material that emits radiation or particles from the nucleus of its atoms.

Dissolved nitrogen

Nitrogen is readily dissolved in body fluids. The greater the pressure, the more gas is dissolved.

At high altitudes, the pressure of the atmosphere is much less than at ground level and aircraft have their cabins pressurised to provide a comfortable environment. Not only is it essential to provide enough oxygen for breathing, but without this pressure, the levels of nitrogen dissolved in the blood would change quickly, giving rise, as pressures fall, to bubbles of nitrogen forming in the blood and causing the effect that scuba divers call "the bends", which in flying is known as aeroembolism.

▲ Auroras are constantly changing glowing lights seen in the night sky, mainly within the Arctic and Antarctic Circles. Auroras result from an interaction between electrons flowing through space and gases in the upper atmosphere. The auroras may occur as veils, curtains and other shapes. A number of colours are associated with auroras, each reflecting the gas that is being bombarded with electrons. Red colours, produced by the bombarding of nitrogen gas, normally fringe the green colours produced by bombardment of oxygen atoms. This picture shows the red fringe of an aurora as seen from space.

◄ Nitrogen gas is converted to soluble nitrogen dioxide by lightning.

EQUATION: Formation of nitrogen dioxide during lightning flashes

Stage 1: Nitrogen + oxygen ⇨ nitric oxide

$$N_2(g) + O_2(g) \Rightarrow 2NO(g)$$

Stage 2: Nitric oxide + oxygen ⇨ nitrogen dioxide

$$2NO(g) + O_2(g) \Rightarrow 2NO_2(g)$$

EQUATION: Formation of natural acid raindrops after lightning

Nitrogen dioxide + water ⇨ nitric acid + nitric oxide (which can react with more oxygen as above)

$$3NO_2(g) + H_2O(l) \Rightarrow 2HNO_3(aq) + NO(g)$$

"Refining" nitrogen

Nitrogen at normal air pressure and temperature contains so much energy that it is a gas. Nitrogen makes up about three-quarters of the mixture of gases in the atmosphere. However, nitrogen is never found in nature as a liquid or a solid.

Liquid nitrogen

To make nitrogen into a liquid the gas has to be deprived of energy by cooling. However, this is not very easy because the temperature at which nitrogen liquifies is -196°C.

To make liquid nitrogen easier to obtain, air is cooled while under pressure. You can think of the process as being rather like an air "refinery". Each of the gases in the air has its own boiling point (where it changes from gas to liquid). As this is reached, the element liquifies. Liquid oxygen forms at -183°C. Nitrogen has the lowest boiling point of any of the gases in the air and it is the last to be collected.

▼ The composition of clean air.

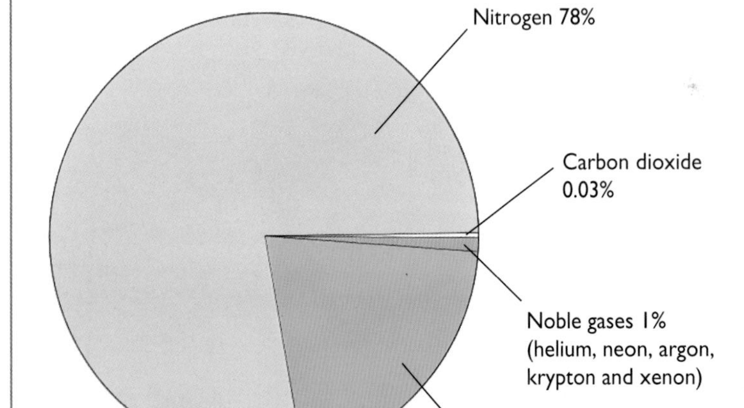

Distilling nitrogen

Often the easiest way to obtain nitrogen is to liquify air in bulk and then allow the oxygen to boil off. This process is called distillation, and it is a very low temperature version of, for example, how oil and other mixtures are obtained from hydrocarbons. The process takes place in a tall column (called a fractionation column) made of a set of "leaky trays". Liquid air is poured onto the trays. There is a boiler at the bottom and a condenser (a trap for the gas given off) at the top.

As the air begins to evaporate, nitrogen is given off as a gas, while the oxygen (which goes into a gas less easily than nitrogen – it is less volatile) drips to the bottom. In this way carefully controlled conditions can separate out the nitrogen from the oxygen.

The liquid air is kept under a pressure of five atmospheres in the tower, at which pressure the boiling point of the nitrogen and oxygen are much higher (and so easier to maintain) than at normal atmospheric pressure.

volatile: readily forms a gas.

▶ Fractional distillation of air. The liquid air is pumped into a fractionating column and subjected to pressure and temperature changes to separate the component gases that make up air.

Air is cooled to -190°C and pressurised to generate liquid air.

Nitrogen gas. Nitrogen boils at -196°C. It is used to make fertilisers and nitric acid.

Some liquid oxygen is used to cool the nitrogen and liquify it. This is possible because the nitrogen is at a higher pressure than the oxygen and so has a higher boiling point.

Argon gas. Argon boils at -186°C and is used as an unreactive filling for some light bulbs.

Liquid oxygen. Oxygen boils at -183°C and is used for breathing apparatus.

▲ Nitrogen has many industrial applications. It creates an inert environment for storing experimental samples during research.

Why nitrogen is liquified under pressure

Chemists make use of the Gas Laws when they are separating out nitrogen from other atmospheric gases. These laws describe how a gas behaves, and are one of the cornerstones of science.

To make nitrogen into a liquid, where the volume occupied by the molecules is much smaller than a gas, the gas volume has to be reduced.

One of the Gas Laws states:

The volume of a gas reduces as the temperature is *lowered* and/or if the pressure is *increased*.

This law means that chemists can work with a combination of lowering the temperature and raising the pressure to find the most economical way of liquifying the gas.

▲ Liquid nitrogen boiling vigorously at room temperature

Uses for inert nitrogen

For many purposes nitrogen can be thought of as an inert gas. This means that it does not react easily with other gases or liquids.

At first this might seem to be a very unpromising property for a gas, but in fact it has found many applications, in food packaging, food preservation and light bulbs.

In each case nitrogen is used *because* it doesn't react. For example, nitrogen gas is used in some forms of food processing to prevent oxygen getting to the food and causing it to spoil. Nitrogen gas in light bulbs is designed to make the filaments last much longer.

Preserving food freshness

Food spoils because of the presence of oxygen (which reacts with the food to produce new, less tasty, chemical compounds). One way of preserving the freshness of food therefore is to fill the space above the food with nitrogen, which replaces the oxygen and does not react with the food at all. (Sometimes nitrogen is called an antioxidant, i.e. it keeps food from oxidising.)

Food can be frozen at very low temperatures using nitrogen. During the freezing process, the food does not have time to spoil as it would when frozen slowly in air.

Other foods that are prepared by roasting could be severely altered by the presence of oxygen in the air. For example, the flavour of roasting coffee will be lost in air, so coffee is roasted in an oven containing nitrogen.

Some snack foods are filled with nitrogen for this reason. Look for the inflated foil bags of crisps and other snacks.

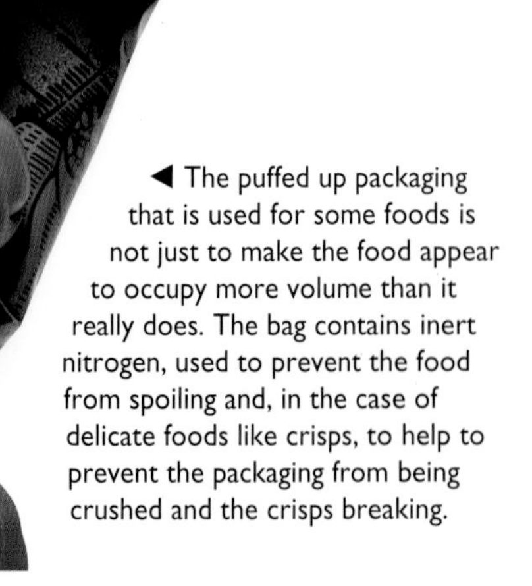

◀ The puffed up packaging that is used for some foods is not just to make the food appear to occupy more volume than it really does. The bag contains inert nitrogen, used to prevent the food from spoiling and, in the case of delicate foods like crisps, to help to prevent the packaging from being crushed and the crisps breaking.

antioxidant: a substance that prevents oxidation of some other substance.

How a light bulb works

Light bulbs (incandescent bulbs) rely on the light given out by a filament as it gets very hot. The first attempts to make an incandescent bulb were a failure because the filament was left in air, where it reacted with the oxygen and burned away within seconds.

To try to solve this problem the filaments were put in a glass bulb from which as much air as possible had been withdrawn. But this was not very successful either. A vacuum is difficult to produce, and should a bulb break, the resulting implosion of the glass can cause possible harm to anyone standing nearby. But even if it didn't break, the glass quickly blackened and so became unusable. This is because the hot filament sends out a stream of particles, gradually wasting away (and thus producing a short light bulb life). Particles of metal were deposited on the inside of the glass bulb, thus darkening it and allowing less useful light to get out.

Both of these disadvantages were overcome using nitrogen gas. By filling the bulb with nitrogen, oxygen is automatically pushed out. This means that the filament will not burn away. At the same time the presence of gas inside the bulb means that a low pressure is not needed and the bulb won't implode if it breaks.

The nitrogen also helps to give the filament a longer life. As particles (atoms) of metal try to leave the hot filament, they find themselves surrounded by molecules of nitrogen gas. The result is there is a much greater chance that the metal atoms will simply bounce back on to the filament again. Not only does this help preserve the filament, but it also means that fewer atoms reach the glass bulb to darken it.

Ammonia

Ammonia, whose formula is NH_3, is a compound of nitrogen and hydrogen. The word may come from an association with the ancient Egyptian god Ammon. In former times ammonia was obtained by boiling animal horns and hoofs.

Ammonia is a colourless gas. It is not actually poisonous, but inhaling its fumes can cause people to stop breathing, so it can have very serious effects.

Ammonia can always be detected by its irritating smell. It is famous for its use as "smelling salts" in the last century, in which a compound of ammonia, ammonium carbonate, was used. Smelling salts were used to revive women who had swooned (not actually fainted, but who were putting on a show as a social gesture that they had been offended). It was very difficult to continue to pretend to have fainted when the smell of ammonia was irritating the throat, and so they "revived".

Ammonia is extremely soluble in water, where it produces an alkaline liquid, ammonium hydroxide. This dilute solution is commonly sold for use in household disinfectants and cleaning under the title household ammonia.

Ammonia is also one of the most important starting materials for making artificial fertilisers.

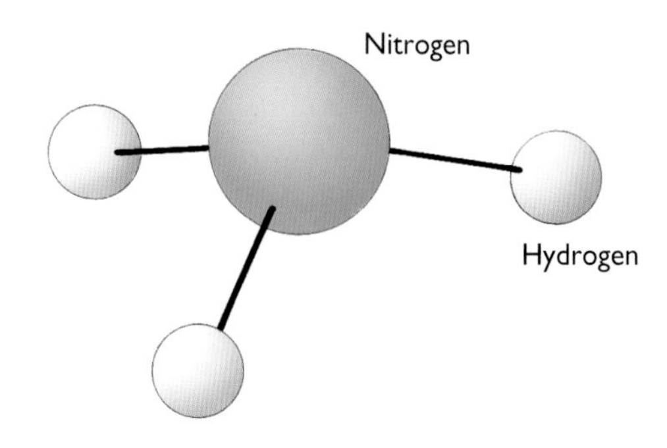

▲ A representation of an ammonia molecule.

Ammonia for refrigeration

Ammonia gas has a high boiling point (about -33°C). It also absorbs a large amount of heat energy as it vaporises. It is not corrosive and is easy to obtain. This makes it particularly suitable for use in refrigeration systems, which work by boiling and then liquifying a material as it passes through the refrigeration pipework.

► Ammonia can be added to irrigation water and applied to fields at the same time as they are watered. In such diluted amounts its smell cannot be detected.

alkaline: the opposite of acidic. Alkalis are bases that dissolve, and alkaline materials are called basic materials. Solutions of alkalis have a pH greater than 7.0 because they contain relatively few hydrogen ions.

resin: natural or synthetic polymers that can be moulded into solid objects or spun into thread.

soluble: a substance that will readily dissolve in a solvent.

◀ The TransAlaskan pipeline uses refrigeration by ammonia in the thousands of legs that support the pipeline over permafrost areas. The refrigerated legs are needed to prevent the permafrost from melting, allowing the legs to founder and causing the pipeline to fracture.

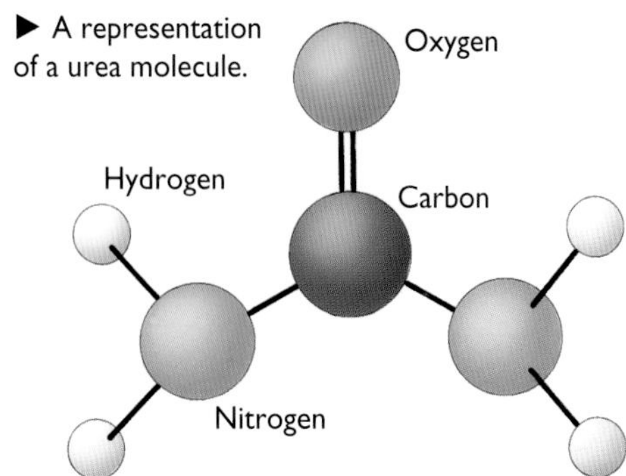

▶ A representation of a urea molecule.

Uses of ammonia

About three-quarters of the ammonia produced is used to make fertilisers. It can be sprayed on crops, on the ground or injected as a gas into the soil. Because it is so soluble it can also be readily added to irrigation water.

Ammonia is also a starting material for a substance called urea, $CO(NH_2)_2$, which is used as a fertiliser in its own right and also in the manufacture of resins, some cosmetics, and medicines.

If ammonia is not used directly as a fertiliser, it is mainly converted to ammonium nitrate, by reacting ammonia with nitric acid. Ammonium nitrate can also be used in explosives such as TNT and dynamite (see page 27).

Ammonium chloride is used as a cleaning agent for metals prior to their being painted, galvanised or soldered. Ammonium chloride is also found in the paste inside all zinc–carbon dry cells.

Ammonia is also used by some craftspeople because, when soaked into wood, it makes the wood behave as a plastic which can then be shaped. As the ammonia evaporates, the wood again becomes hard.

EQUATION: Reaction of ammonia and nitric acid

Ammonia + nitric acid ⇨ ammonium nitrate

$NH_3(aq) + HNO_3(aq) \Rightarrow NH_4NO_3(aq)$

Soluble alkaline ammonia

Ammonia is an alkaline substance, so it will cause a colour change in a chemical indicator. Ammonia is also extremely soluble and one volume of water will dissolve four hundred times its own volume of ammonia. The density of the solution falls to 0.880 grams per millilitre as it becomes saturated and the solution is known as "880 ammonia".

Both of these properties of ammonia can be demonstrated in a dramatic way, as shown on this page, in a classic laboratory experiment.

Preparing dry ammonia in the laboratory

Ammonium sulphate and calcium hydroxide are ground up and mixed using a pestle and mortar. The mixture is put into a long-necked, round-bottomed flask, and the neck clamped so that it points downwards.

The mixture is heated. The heat energy causes a reaction and ammonia and water vapour are given off. The (wet) ammonia gas is passed through a vertical column (called a drying tower) containing calcium oxide (another base, so no reaction occurs). The calcium oxide (lime) takes up the water from the ammonia (to make calcium hydroxide, slaked lime) and the dry ammonia gas now passes to an inverted collecting flask. Here it displaces air from the flask because ammonia has a lower density than air.

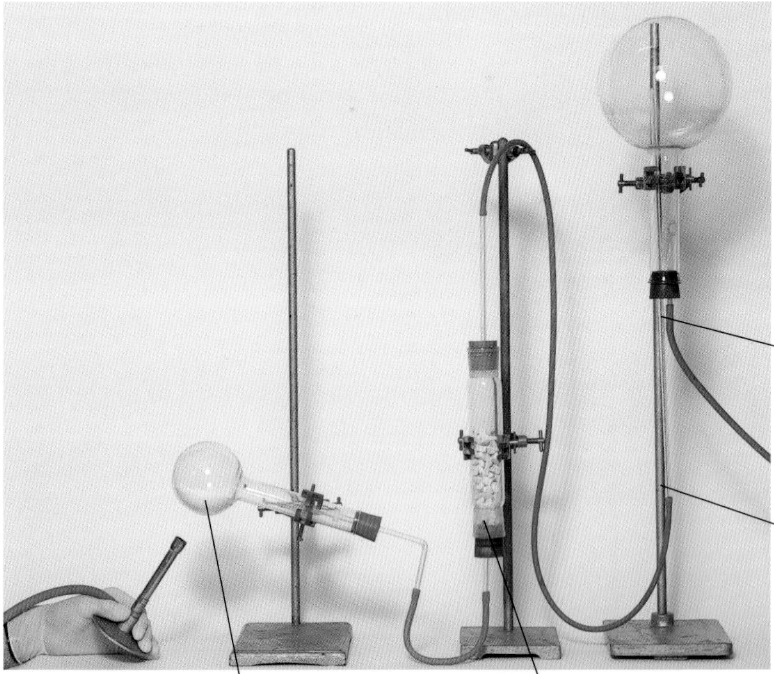

Short exit tube

Long entry tube

Ammonium sulphate and calcium hydroxide mixture.

Calcium oxide drying tower.

◀ Laboratory apparatus for preparing dry ammonia.

EQUATION: Producing ammonia in the laboratory

Ammonium sulphate + calcium hydroxide ⇨ ammonia + water + calcium sulphate

$(NH_4)_2SO_4(s) + Ca(OH)_2(s) \Rightarrow 2NH_3(g) + 2H_2O(l) + CaSO_4(s)$

An ammonia fountain

❶ The preparation apparatus is removed, the short exit tube is blocked and the long entry tube is led from the flask to a trough of water containing a chemical indicator. The indicator is colourless because the water is neutral.

❷ Ether liquid is poured on to the flask, causing the temperature of the gas to drop. The gas therefore contracts, and the pressure inside the flask is reduced.

Air pressure is now able to push water from the trough up the thin tube and into the flask.

As soon as the first drops of water come into contact with the ammonia gas, the ammonia goes into solution and there is a rapid drop of pressure in the flask. Air pressure then causes the water to spurt into the flask, producing the fountain.

❸ and ❹ The flask fills with bright pink water, as the ammonia turns the water alkaline and the chemical indicator (phenolphthalein) turns pink.

alkaline: the opposite of acidic. Alkalis are bases that dissolve, and alkaline materials are called basic materials. Solutions of alkalis have a pH greater than 7.0 because they contain relatively few hydrogen ions.

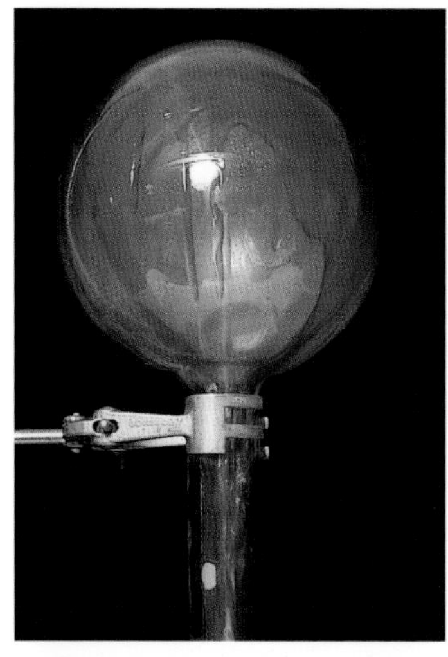

▲❷ The ammonia fountain begins.

▼❶ The apparatus before the ammonia fountain.

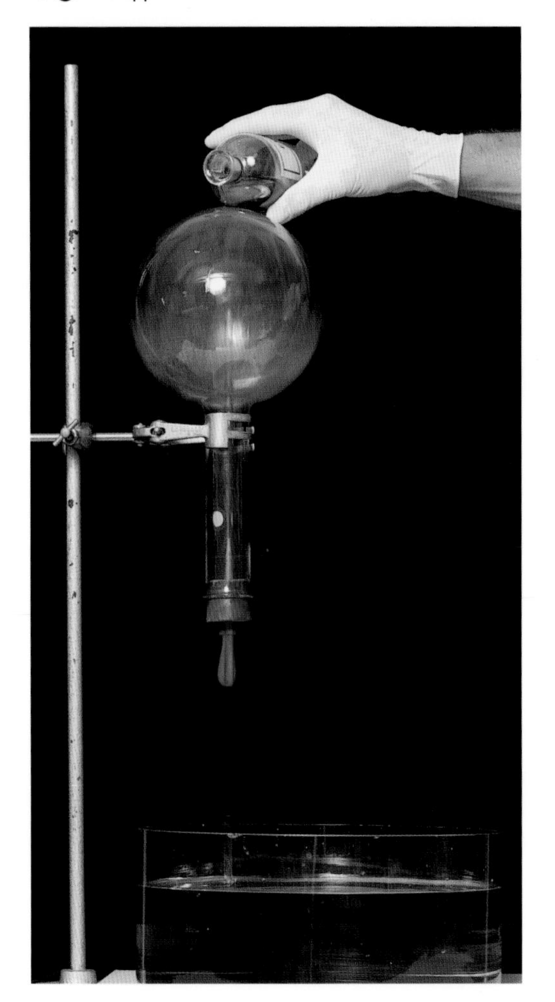

▼►❸ and ❹ The pink colour of the liquid is the effect of the indicator showing the presence of an alkaline solution.

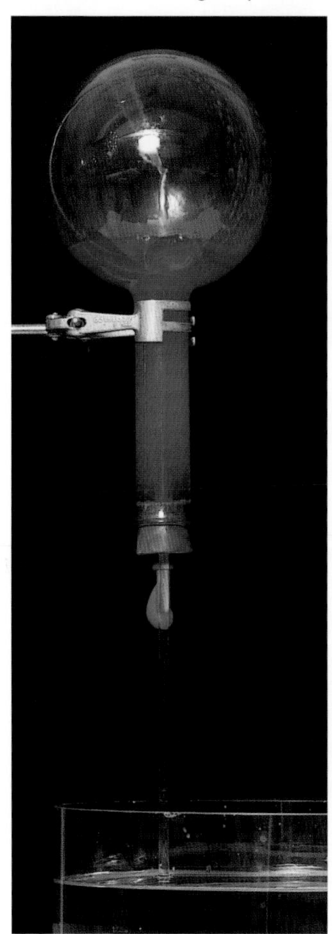

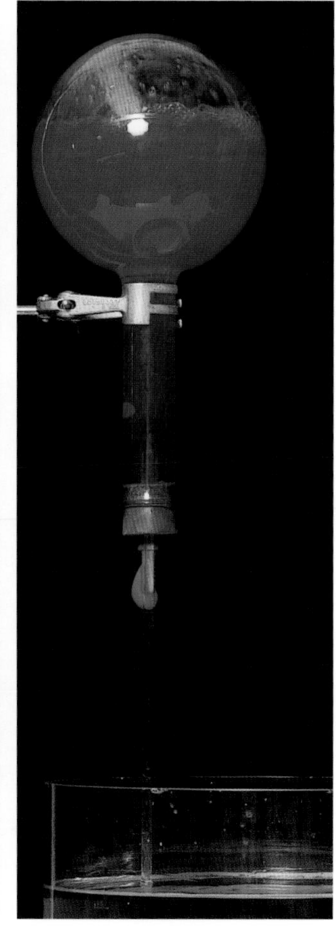

Making ammonia industrially

The industrial method of making ammonia is called the Haber–Bosch process after its inventors. In the process, nitrogen and hydrogen gases are reacted at high temperatures (up to 600°C) and high pressures (up to 600 atmospheres). The reaction works best if this is done in the presence of iron, although iron is a catalyst in the reaction and so remains unchanged.

The ammonia produced is liquified to separate it from the unreacted gases, which are then recycled.

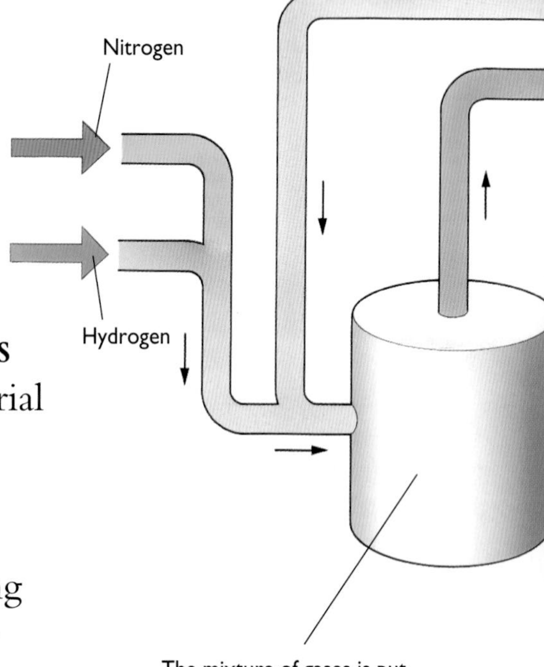

The mixture of gases is put under great pressure and heated to a high temperature.

Limitations of the Haber–Bosch process

This process has produced the foundation material for important fertilisers and has thus greatly contributed to the increase in the world's food supply. However, the process consumes large amounts of energy and the equipment, operating at high temperatures and pressures, is expensive to make.

The source of hydrogen from the reaction is mainly petroleum, a nonrenewable fossil fuel. For this reason, fertilisers are expensive and scientists are looking for a natural way to fix nitrogen for use as a fertiliser.

EQUATION: Production of ammonia by the Haber-Bosch process

Hydrogen+ nitrogen ⇨ ammonia

$3H_2(g) \quad + \quad N_2(g) \quad \Rightarrow \quad 2NH_3(g)$

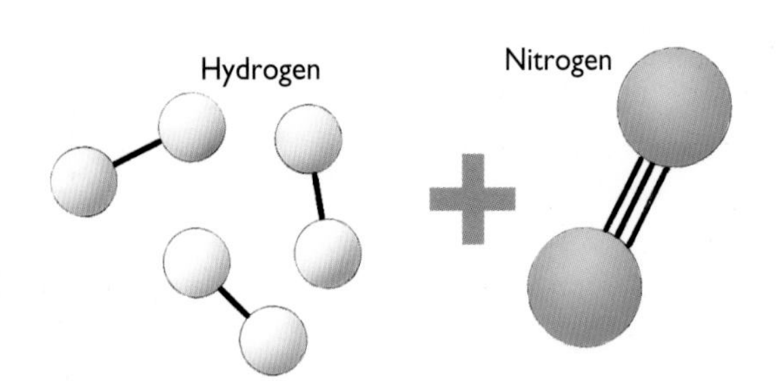

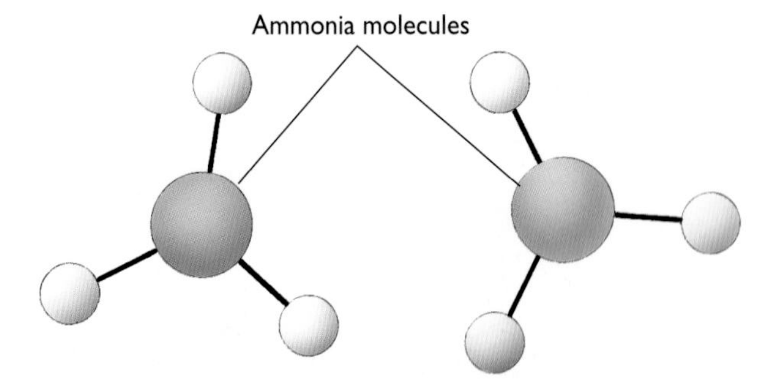

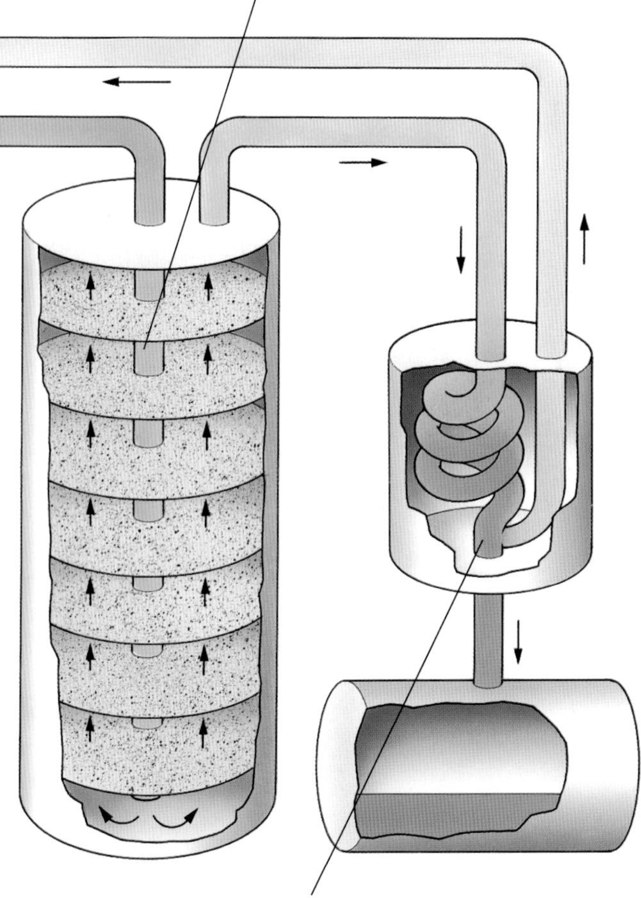

catalyst: a substance that speeds up a chemical reaction but itself remains unaltered at the end of the reaction.

fixing: making solid and liquid nitrogen-containing compounds from nitrogen gas. The compounds that are formed can be used as fertilisers.

Also... Ammonia as a source of hydrogen

Hydrogen is a highly inflammable gas that is dangerous to transport. To solve this problem, hydrogen is often transported in the form of safe liquid ammonia. When it reaches its destination, the ammonia is heated, and decomposes to one-quarter nitrogen and three-quarters hydrogen.

◄▲ The catalytic towers in an ammonia plant.

Naturally fixed nitrogen

Nitrogen can be made in a chemical plant for use as a fertiliser. However, there are natural processes that perform a much more efficient job of "fixing" nitrogen into a form that is available for plant use.

Several species of bacteria, fungi, and blue–green algae are able to "fix" nitrogen. They convert nitrogen into ammonia, a soluble material that can be absorbed by plants.

One of the most common nitrogen-fixing group of bacteria is called *Rhizobium*. These bacteria make the nodules, or galls, on the roots of legumes (plants like clover, beans, peanuts and alfalfa).

Both the bacteria and the plants benefit from this relationship. The bacteria need the plants for food; the waste products of the bacteria are nitrogen compounds, which are important to plant growth.

Natural nitrogen-producers are an efficient method of applying nitrogen fertiliser to a field. This is why, for several centuries, legumes have been grown alternately (in rotation) with "nitrogen-hungry" crops like wheat and barley.

Nitrogen-fixing bacteria can also be used to improve the yields of water-loving plants such as rice. In the flooded rice paddy fields the nitrogen is fixed by blue–green algae that live together with species of water fern. By planting this fern in rice fields, nitrogen compounds are fixed and released into the water where they can be absorbed by the roots of rice plants.

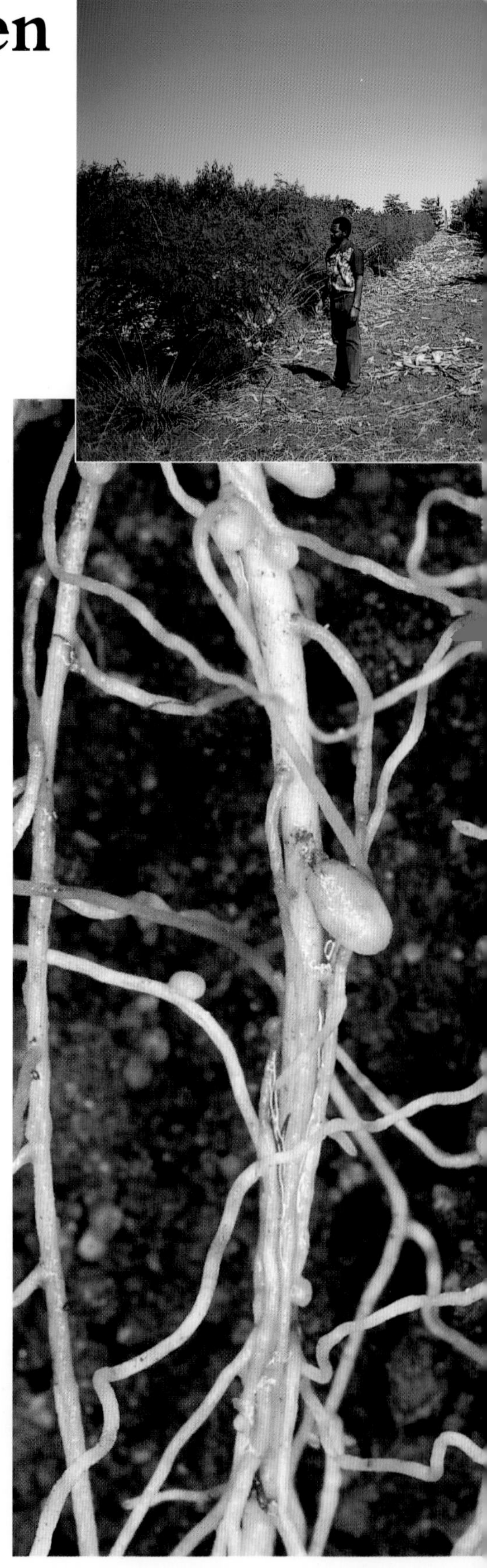

◀ A wide range of leguminous plants, for example these tropical acacias, exists in the world.

fixing: making solid and liquid nitrogen-containing compounds from nitrogen gas. The compounds that are formed can be used as fertilisers.

▶ Blue–green algae in the water that floods a paddy field are able to fix nitrogen.

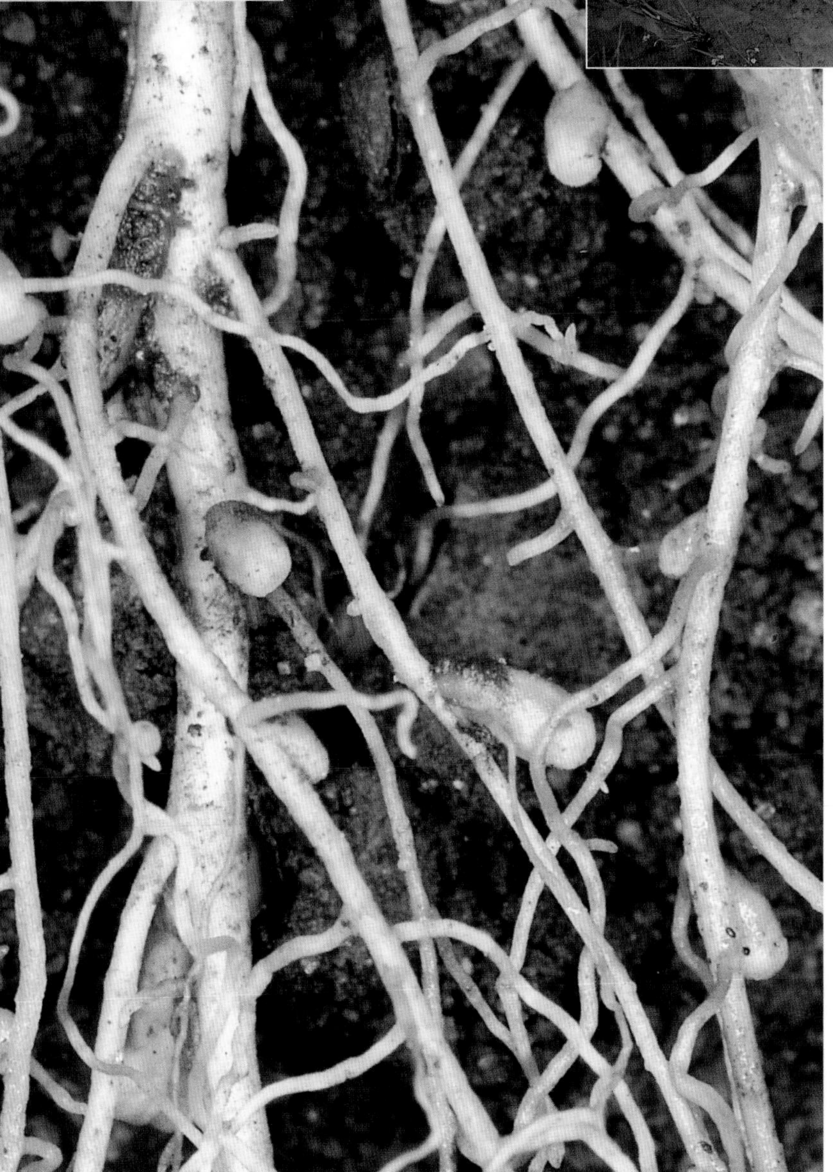

▼ Nodules on a nitrogen-fixing legume.

The importance of natural sources of fixed nitrogen

About one kilogram of nitrogen may be fixed (made into nitrogen-containing compounds) per hectare of the Earth's surface per year by lightning flashes.

However, the majority of nitrogen is fixed in the soil. Bacteria in legume nodules extract about fifty kilograms of nitrogen per hectare from air in the soil, while other forms of bacteria not connected to legumes fix a further five kilograms per hectare per year.

Yet despite these seemingly large amounts of nitrogen fixed from the air, they only make up about one-third of the total requirements of growing plants, and the remainder must come from recycling nitrogen as compost and manure or from artificial supplies.

Nitrogen-based fertilisers

Plants obtain their nitrogen as soluble nitrate compounds. Nitrogen-based fertilisers (in the form of soluble nitrate) are among the most important of all aids to producing more crops for the world to eat.

Using manure is the natural ("organic") way of applying nitrogen fertiliser in addition to that in rainfall or produced by legumes and soil bacteria. Using manure is a way of recycling waste products, and an added advantage is that manure contains a wide variety of elements in addition to nitrogen that can be used for the benefit of plants and the soil.

Manure tends to act slowly, relying on bacteria to break down the waste organic matter. As part of the process of breakdown, bacteria release nitrogen as ammonia and soluble nitrates.

For more immediate, as well as sometimes more convenient, application, concentrated nitrogen-containing fertilisers, obtained from nitrate rock or produced from ammonia, are applied to a soil.

Applying nitrogen fertiliser

There are several methods of applying nitrogen fertiliser. On highly mechanised farms it may be applied in the form of an ammonia solution in water that is injected about 15 cm under the soil's surface to prevent the liquid from vaporising. More commonly, a solution containing ammonia compounds is sprayed onto the fields.

To make the number of applications less frequent, fertiliser can be applied in pellet form. The pellets dissolve and release nitrates throughout a growing season.

Artificial fertilisers do have limitations, however, and not just because they may pollute rivers and lakes. The rapid growth that can occur with some very concentrated formulations can cause the crop to have less taste, be "watery" and have other less desirable qualities. For these reasons, research continues into how to make more effective use of natural nitrogen fixation, producing genetic engineering that may allow all plants in the future to fix their own nitrogen.

► A stick containing nitrate pushed into the soil of a house-plant.

▲ Potassium nitrate occurs naturally as the mineral saltpetre, which is found in abundance in Chile and in much smaller deposits as a white crust in rocks and caves.

Nitrate-containing rocks

Rocks containing nitrogen are uncommon. They consist mainly of a rock called saltpetre (potassium nitrate), the largest source of which is in the deserts of Chile. Because natural supplies of nitrates are so scarce, and the need for fertilisers so important, countries are reluctant to be dependent on such overseas sources. As a result, the Haber–Bosch process of using ammonia (see page 16) has been adopted throughout the world.

▶ The majority of ammonium nitrate is produced from ammonia using the Haber–Bosch process.

Nitrogen cycle

Nitrogen is continually transferred between the land, animals, plants and the atmosphere in a never-ending cycle. This nitrogen cycle is one of the most important cycles for life on Earth. Here are the main stages summarised from the detail provided on the previous pages:

❶ The cycle begins when lightning flashes provide the energy for atmospheric nitrogen and oxygen to combine to form dilute nitric acid.

❷ Several species of bacteria, fungi, and blue–green algae can convert nitrogen into nitrogen-containing compounds such as nitrates. These compounds are then used in photosynthesis to make new sources of energy for the growth of plant tissue.

❸ Along the food chain, animals use the nitrogen locked up in plant and animal tissues as a way of getting the nitrogen they need to build new cells. The nitrogen is obtained during the process of digestion.

❹ All animals produce waste nitrogen in the form of a nitrogen-containing compound called urea. This is removed from the body during excretion. It will either evaporate or be broken down by bacteria. Nitrogen-containing compounds are very soluble and rainwater may wash them into rivers, where they may cause pollution problems.

❺ Bacteria involved in decomposing waste and dead material release nitrogen from its compounds, forming gas that can escape to the atmosphere, thus completing the cycle.

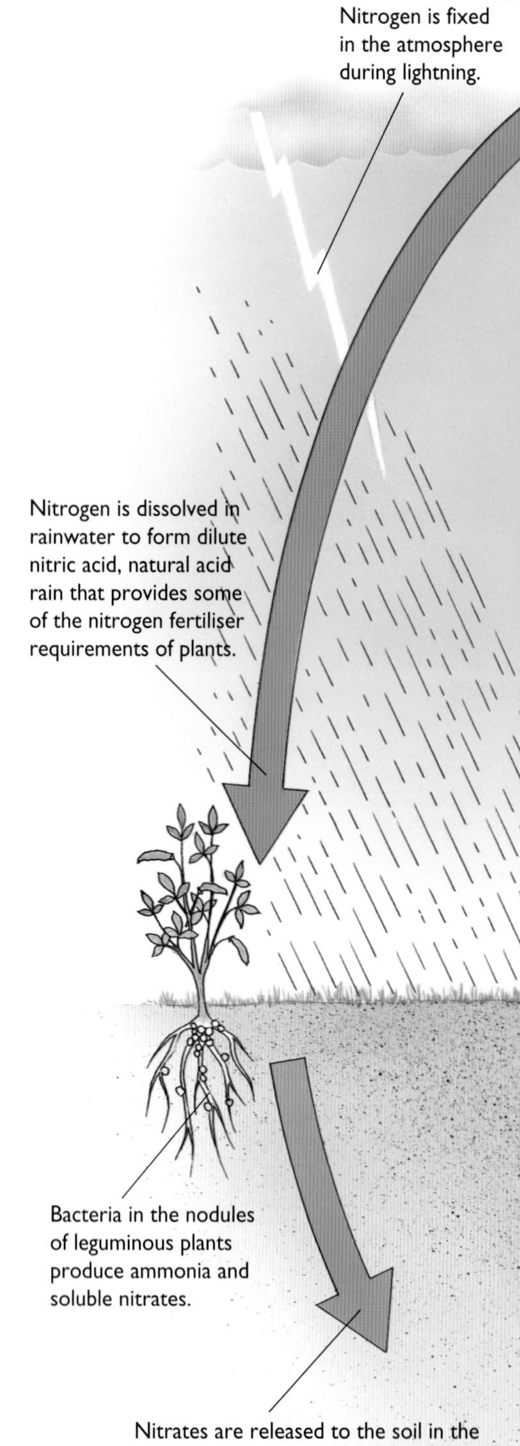

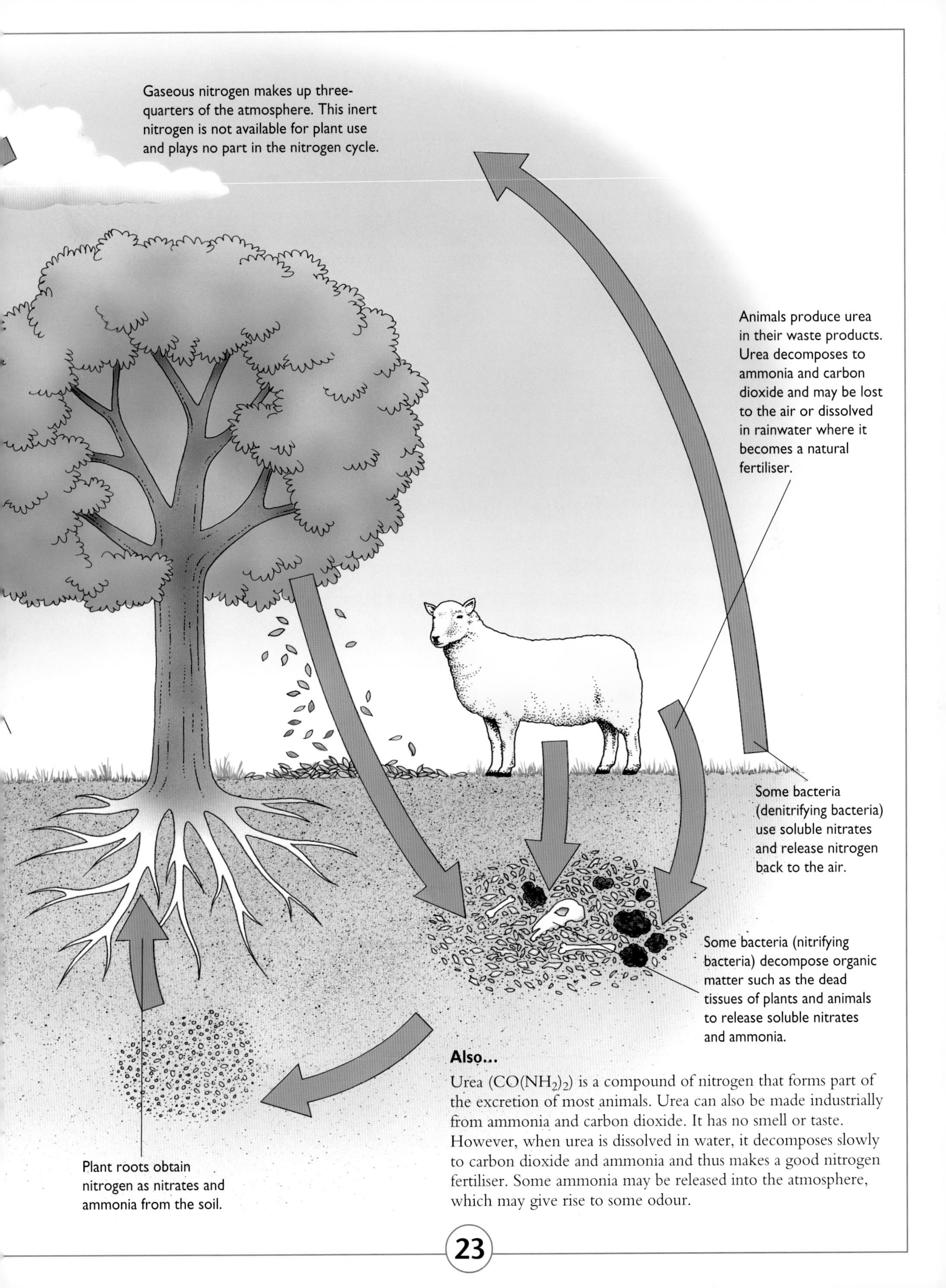

Also...

Urea ($CO(NH_2)_2$) is a compound of nitrogen that forms part of the excretion of most animals. Urea can also be made industrially from ammonia and carbon dioxide. It has no smell or taste. However, when urea is dissolved in water, it decomposes slowly to carbon dioxide and ammonia and thus makes a good nitrogen fertiliser. Some ammonia may be released into the atmosphere, which may give rise to some odour.

Demonstrating nitrogen-based explosives

Solids and liquids occupy very little volume compared with the same number of molecules of a compound as a gas. The rapid change from solid or liquid to gas is the key to creating an explosion.

An explosive is a substance that is normally unreactive, but which can be made to react very violently, changing from a solid or a liquid to a gas. The explosion is created because of the pressure of the expanding gas on surrounding solid or liquid materials.

Many explosive materials simply burn when they are not contained in some form of packing. Thus, for example, gunpowder can be laid as a trail of powder to make a burning fuse, but if the trail leads to gunpowder confined in a barrel, the gunpowder will explode. This shows that the process of burning is quite slow, and that a container is needed so that pressure can build up to give an explosive effect.

Many materials can be made explosive. For example, petrol vapours can explode, as can coal dust and even flour dust. Thus, providing a material with a large amount of surface area – as happens with powders – also helps to create an explosion.

▼ The relative space occupied by a solid, liquid and a gas helps us to understand the effect of explosives. A sudden change of state of a substance from a solid to a gas creates a rapid expansion and consequent displacement of whatever material is around that substance.

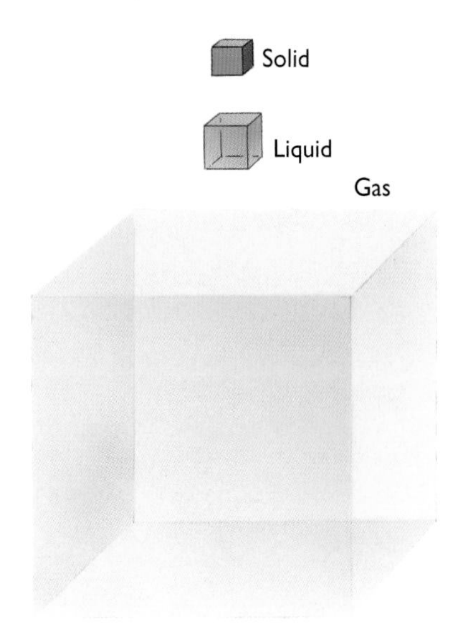

▶ Orange ammonium dichromate crystals are dense and occupy relatively little space. During the reaction shown, nitrogen is released as a gas, leaving green chromium oxide behind.

Ammonium dichromate

Boiling tube cannon

In the demonstration on this page some crystals of orange ammonium dichromate have been strongly heated in a boiling tube. This adds heat energy to the crystals. Notice that the boiling tube is heated at the open end. This means that the reaction first occurs in the crystals near the open end, and only later to those near the closed end.

The first change is that the crystals glow red hot. At this point, the external heating is removed. The chemical decomposition gives out heat (as nitrogen molecules form), which causes the reaction to spread along the mixture. The crystals decompose to give nitrogen gas, water (as steam) and a bulky green solid called chromium oxide.

As the reaction continues, the chromium oxide blocks the open end of the tube, so as the heat energy is used to create nitrogen gas, pressure builds up towards the closed end of the tube until eventually the gas pressure is enough to shoot the chromium oxide out as fragments. This explosive violence is similar to the way ash is shot out from a volcano, or cannonballs, grapeshot or shells fly out of a piece of artillery. In every case it is the rapid creation of nitrogen gas that builds up pressure and eventually breaks apart any nearby solids.

decompose: to break down a substance (for example by heat or with the aid of a catalyst) into simpler components. In such a chemical reaction only one substance is involved.

explosive: a substance which, when a shock is applied to it, decomposes very rapidly, releasing a very large amount of heat and creating a large volume of gases as a shock wave.

▼ Fine particles of chromium oxide are ejected from the boiling tube cannon.

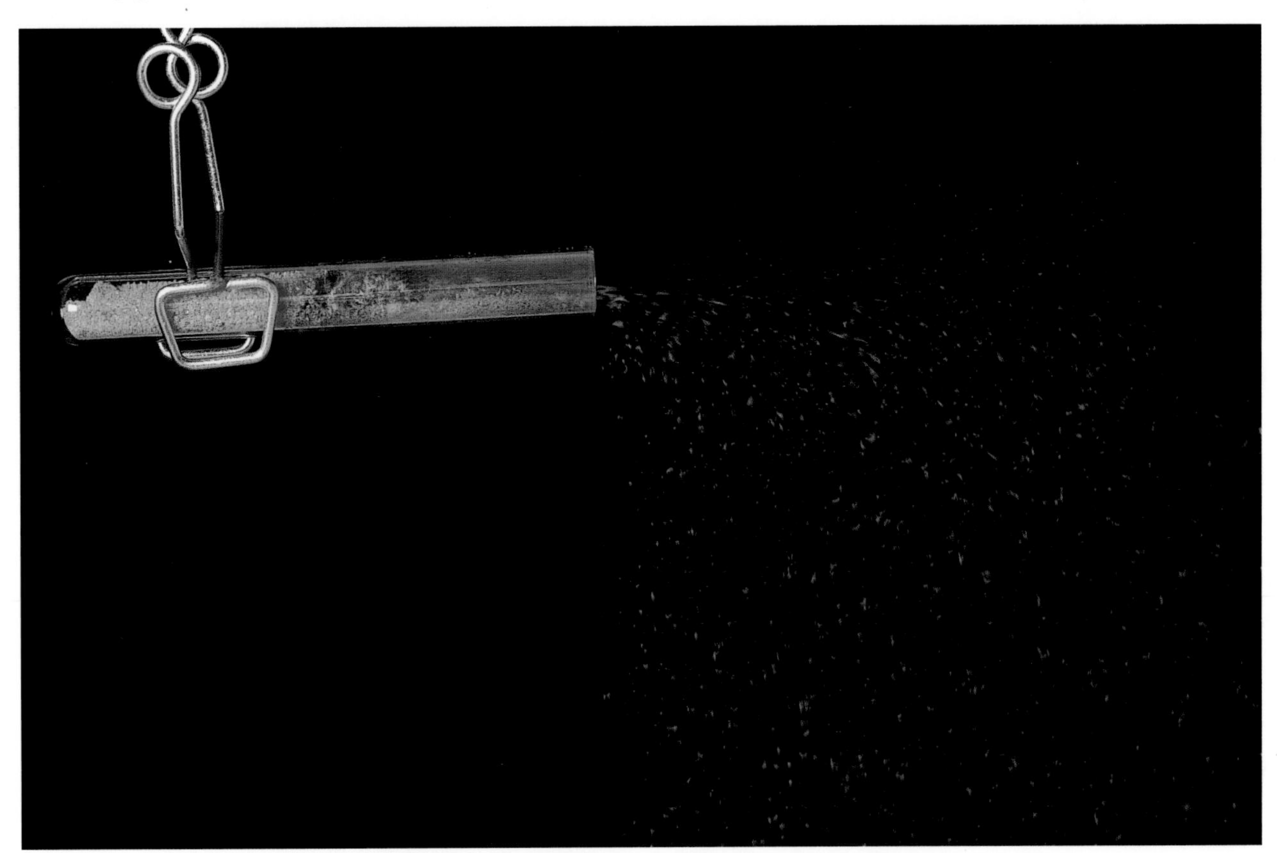

EQUATION: Heating ammonium dichromate

Ammonium dichromate ➪ nitrogen + chromium oxide + water

$(NH_4)_2Cr_2O_7(s)$ ➪ $N_2(g)$ + $Cr_2O_3(s)$ + $4H_2O(l)$

Nitrogen-containing explosives

Explosives design is a very specialised form of chemistry. Explosives can be used for propelling bullets and shells for use in war, but they can also be used for a wide range of peaceful purposes. Each type of explosive has its own exploding characteristics, described here.

▲ By 1242 Roger Bacon was using a mixture of 41.2 parts saltpetre, 29.4 parts charcoal, and 29.4 parts sulphur for his gunpowder in making fireworks. A gunpowder mixture is still used for fireworks, such as this catherine wheel.

Gunpowder

The world's first true explosive, gunpowder, was invented in China in about 1000 AD. It is a mixture of 10% sulphur, 15% charcoal, and 75% potassium nitrate.

Gunpowder is one of the most important chemicals ever to be invented, because by allowing people to use explosive devices including guns and cannons, it changed the course of history.

Gunpowder was particularly easy to make explode because it is very sensitive to sparks. However, this same property meant that accidents were common. To reduce the chances of accidental explosions, gunpowder was mixed with graphite.

The main uses for gunpowder today are for fireworks and in artillery shells and fuses for hand grenades.

▼ Many guns use explosives derived from nitrogen.

Nitroglycerin and dynamite

Glyceryl trinitrate (called nitroglycerin) is the most unstable of the common explosives. It is a pale yellow oil prepared by reacting nitric acid with glycerol (a product of oil refineries).

Alfred Nobel was the first to make a stable form of nitroglycerin by making it into a solid. In this way he created the world's most widely used high explosive – dynamite.

explosive: a substance which, when a shock is applied to it, decomposes very rapidly, releasing a very large amount of heat and creating a large volume of gases as a shock wave.

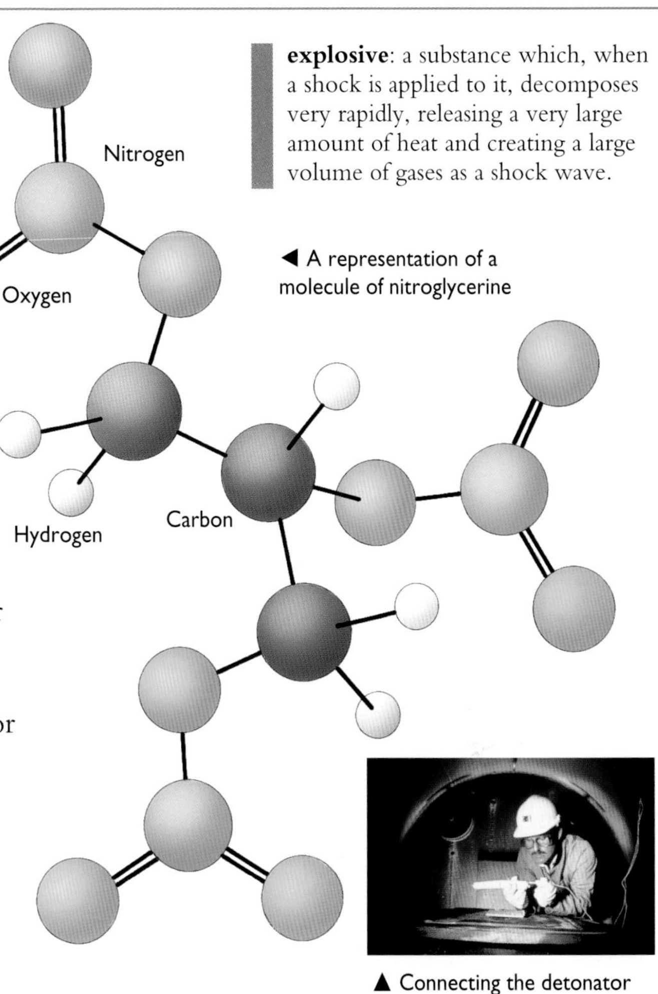

◀ A representation of a molecule of nitroglycerine

▲ Connecting the detonator to an explosive charge.

Ammunition

Ammunition is the word used to describe a wide variety of objects that are propelled to their target. Any piece of ammunition will have a primer, or detonator, a propellant, the material that explodes, and a projectile, such as the shell or bullet. All of this is usually contained in a casing.

Most modern primers are called percussion caps. They contain sensitive explosives that detonate on a shock from, for example, the hammer of a gun.

The main propellant for ammunition is nitrocellulose. This was traditionally called guncotton, because at one time cotton fibres were soaked in liquid explosive. Modern guncotton uses wood fibre.

Projectiles used to be small lead balls. Modern ammunition is clad in brass or copper and is given a more streamlined shape that we recognise as bullets.

TNT

TNT is the abbreviation for trinitrotoluene, the most common high explosive. It is produced by reacting toluene with nitric acid in the presence of sulphuric acid.

The advantage of TNT is that it is not sensitive to shocks and so can be handled without risk of accidental explosion. It can even be burned without exploding.

TNT has a low melting point (about 80°C), so it can be melted with steam and poured into the casings of gun shells.

Detonators

Most high explosives can only be detonated with the shock from another explosive. This is why most explosive devices have a main explosive and a detonator. The detonator, or blasting cap, contains small amounts of more sensitive explosive to set off a larger reaction.

There are several common detonating explosives, such as PETN (pentaerythrite tetranitrate) used in blasting caps. RDX (cyclotrimethylenetrinitramine) produces the highest energy yield and detonation pressure of any common explosive. Mixtures of RDX with plastic materials make "plastic explosive". Ammonium nitrate explodes with a blast that is less violent and sudden than many other explosions. This is used to move an obstacle rather than to shatter it. Amatol (ammonium nitrite and TNT) is used as a blasting charge.

Peaceful explosions

People often associate explosives with death and wanton destruction. However, the vast majority of explosives are used for peaceful purposes, such as to blast rock from quarries or to demolish buildings without damaging the surrounding area.

All explosives decompose to release both gas and heat. It is often thought that, because of this, explosives cause fires. However, explosives are often used to put out fires, such as when burning oil wells have to be capped. Here, the expanding nitrogen gas deprives the burning oil of the oxygen it needs to keep burning and so the flame goes out. At the same time any hot material is thrown clear so it cannot reignite the oil. The oil will then continue to gush, but will not relight.

Explosives used for demolition

Different explosives are suited to different types of building. RDX (see page 27) is the main explosive used for demolishing skyscrapers with steel frames. Dynamite is used to demolish reinforced concrete buildings.

Most demolition experts use relatively small charges placed in holes drilled into the structure. Demolition aims to achieve a progressive collapse of the structure so that it falls in on itself without creating too much dust.

▼ A large road bridge is demolished using explosives.

▲ Modern air bags are filled with sodium azide.

Azide: explosive air bags of nitrogen

No gas cylinder lies behind the life-saving air bag. Everything happens in a flash because of nitrogen chemistry. Nitrogen is produced within one-thousandth of a second, far faster than a gas cylinder could release a supply. Nitrogen is also inert, so it will not contribute to a fire.

Sodium azide is a compound of sodium and nitrogen. All azides are shock sensitive, and lead azide is used as a detonator in some explosives.

Sodium azide is used in air bags in motor vehicles, where it is detonated by an electrical device that is itself triggered by sudden deceleration. Sodium azide decomposes very quickly, releasing nitrogen gas that inflates the air bag.

◀ The hundreds of oil well fires started deliberately by the retreating Iraqi troops during the Gulf War needed to be extinguished quickly because of the environmental damage that was being created. Most were snuffed out using explosives.

Nitrogen dioxide

The three most important gases of nitrogen are nitrogen dioxide (a brown gas that is the cause of the brown part of modern smog), and nitrous and nitric oxides (both colourless gases). All three gases are present in the emissions from vehicles and contribute in various ways to acid rain.

The properties of nitrogen dioxide can be investigated by preparing it in a gas jar in the laboratory or as shown on page 32. Its effects on the environment are discussed on page 34.

❶► Nitrogen dioxide is prepared by pouring concentrated nitric acid on to copper turnings.

❷► As the gas forms, it is a strong brown colour. When it has filled the gas jar it spills over the sides and sinks down the outside of the jar, creating a sinister brown layer on the bench top.

▼ A representation of a molecule of nitrogen dioxide gas.

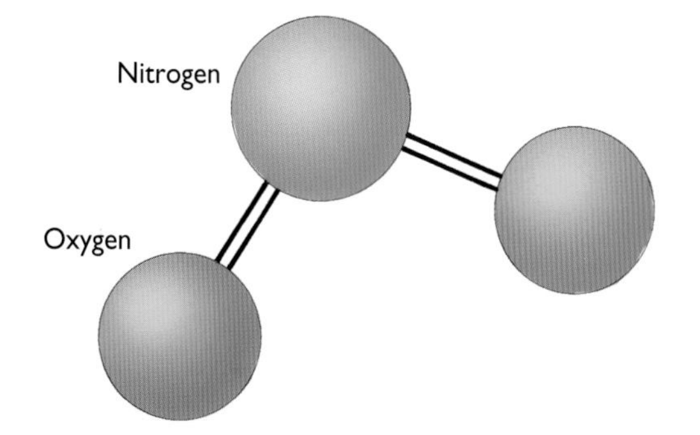

EQUATION: Copper and nitric acid

Copper + nitric acid ⇨ copper nitrate + water + nitrogen dioxide

$Cu(s) + 4HNO_3(conc) \Rightarrow Cu(NO_3)_2(aq) + 2H_2O(l) + 2NO_2(g)$

Also...

Copper is not very reactive, but it will react with concentrated nitric acid, which acts as an oxidising agent rather than an acid.

acid rain: rain that is contaminated by acid gases such as sulphur dioxide and nitrogen oxides released by pollution.

oxidising agent: a substance that removes electrons from another substance (and therefore is itself reduced).

smog: a mixture of smoke and fog. The term is used to describe city fogs in which there is a large proportion of particulate matter (tiny pieces of carbon from exhausts) and also a high concentration of sulphur and nitrogen gases and probably ozone.

❸▼ At the end of the demonstration, water is added to give a blue copper nitrate solution. Nitrogen dioxide is very soluble in water, so the gas jar rapidly clears of the brown gas.

Nitrogen dioxide dissolved in water is nitric acid, one of the two main components of acid rain.

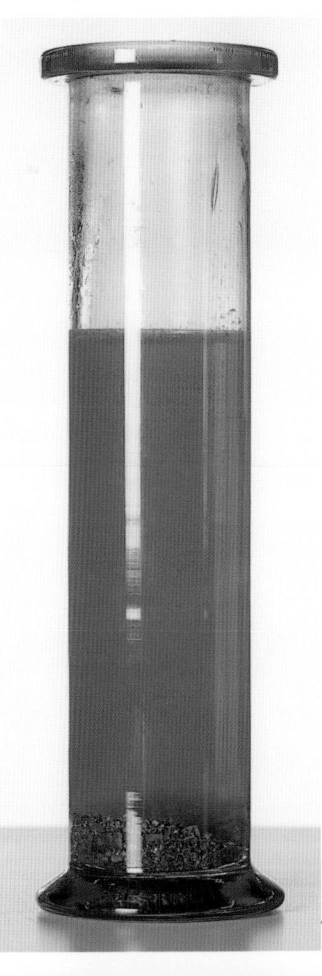

Preparing nitrogen dioxide

This demonstration shows how nitrogen dioxide is produced and what happens if it is cooled.

The demonstration uses lead nitrate as a source of gases. Lead is not a very reactive metal. As a result, its compounds are not very stable and they decompose easily. This is why lead nitrate is a good substance to use for a laboratory demonstration.

In this demonstration, lead nitrate is heated. Heating causes the lead nitrate to decompose to lead oxide. During this process, energy is released, which can be heard as the crystals of lead nitrate break up.

This breaking down process is called decrepitation. It releases both nitrogen oxides and oxygen, as shown by the gas collected at the end of the apparatus.

In the final stage the gases are dissolved in water, creating nitric acid. In the atmosphere nitrogen dioxide would dissolve in raindrops, helping give rise to a natural fertiliser, or, in greater concentrations, to acid rain.

❶ Lead nitrate, a white powder, is heated. It changes colour as it melts and decomposes to lead oxide. The bubbles and the brown fumes are oxygen gas and nitrogen oxides, which will be separated out in step 2.

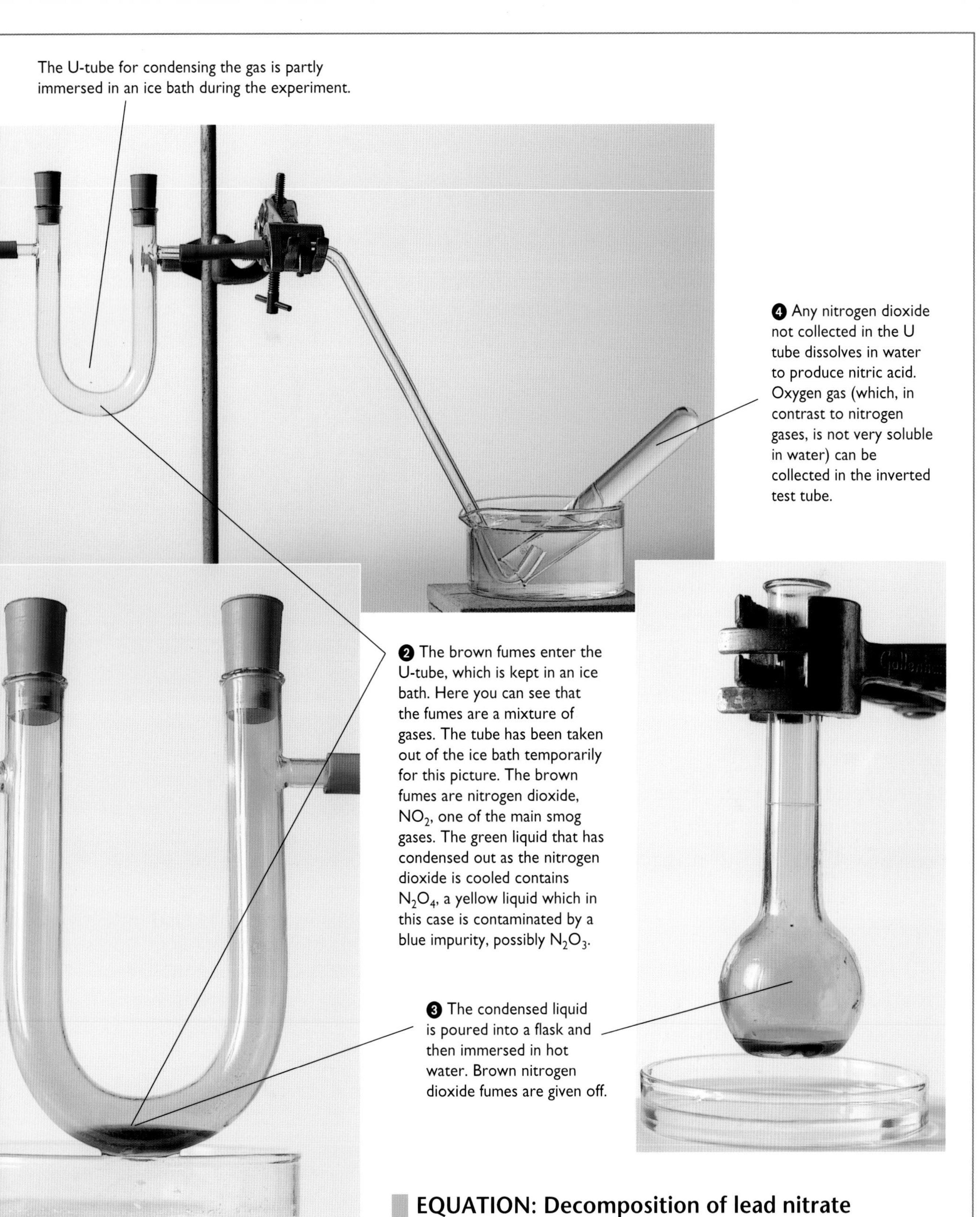

The U-tube for condensing the gas is partly immersed in an ice bath during the experiment.

❹ Any nitrogen dioxide not collected in the U tube dissolves in water to produce nitric acid. Oxygen gas (which, in contrast to nitrogen gases, is not very soluble in water) can be collected in the inverted test tube.

❷ The brown fumes enter the U-tube, which is kept in an ice bath. Here you can see that the fumes are a mixture of gases. The tube has been taken out of the ice bath temporarily for this picture. The brown fumes are nitrogen dioxide, NO_2, one of the main smog gases. The green liquid that has condensed out as the nitrogen dioxide is cooled contains N_2O_4, a yellow liquid which in this case is contaminated by a blue impurity, possibly N_2O_3.

❸ The condensed liquid is poured into a flask and then immersed in hot water. Brown nitrogen dioxide fumes are given off.

EQUATION: Decomposition of lead nitrate

Lead nitrate ⇨ lead oxide + nitrogen dioxide + oxygen

$$2Pb(NO_3)_2(s) \Rightarrow 2PbO(s) + 4NO_2(g) + O_2(g)$$

NO_x

The three important gases of nitrogen, nitrogen dioxide, nitrous oxide and nitric oxide, are collectively referred to as NO_x (pronounced "nox").

Nitrous oxide, or laughing gas, is a colourless gas. When it is inhaled, nitrous oxide makes people feel intoxicated (hence the tendency to laugh). It also makes people lose the sensation of pain and then makes them unconscious. Its effects do not last very long. This is why it is used in dental and other minor operations.

Nitrous oxide dissolves in fatty liquids such as cream. So can be used as the inert propellant in foamed cream.

Nitric oxide is far less pleasant, a colourless gas that reacts on contact with oxygen in the air to produce the brown gas nitrogen dioxide. Nitric oxide is often produced at high temperature and pressure. One of the most common places for this reaction is inside the cylinder of a motor vehicle.

Nitrogen dioxide can damage the breathing passages of the lungs. The damage often goes undetected until days after the gas is inhaled, when it may cause excessive fluid build-up in the lungs.

Nitrogen dioxide is also an important contributor to the production of ozone, which itself has harmful effects. Nitrogen dioxide also contributes to the formation of nitrate particles, another part of smog.

▲ Smog over Los Angeles, California, USA.

Controlling NO_x emissions

It is very difficult to prevent NO_x gases from leaving the exhausts of vehicles, because they are a natural product of internal combustion. Some catalysts can reduce nitric oxide emissions by making nitric oxide oxidise carbon monoxide, producing inert nitrogen, carbon dioxide, and water. However, the catalytic converter in modern vehicles is extremely expensive.

Also... Smog

Smog is a mixture of gases and small particles suspended in the air. The gases come from power station, factory and vehicle exhausts. The main gas produced is nitrogen dioxide, which is responsible for the brown haze that occurs in the sky over many cities. Local visibility is reduced not by the gases, but by small particles in the air, some of them dust and carbon particles, some of them formed as the nitrogen and sulphur gases react to produce sulphates and nitrates.

catalyst: a substance that speeds up a chemical reaction but itself remains unaltered at the end of the reaction.

ozone: a form of oxygen whose molecules contain three atoms of oxygen. Ozone is regarded as a beneficial gas when high in the atmosphere because it blocks ultraviolet rays. It is a harmful gas when breathed in, so low level ozone, which is produced as part of city smog, is regarded as a form of pollution. The ozone layer is the uppermost part of the stratosphere.

smog: a mixture of smoke and fog. The term is used to describe city fogs in which there is a large proportion of particulate matter (tiny pieces of carbon from exhausts) and also a high concentration of sulphur and nitrogen gases and probably ozone.

EQUATION: Reaction of nitric oxide and air

Nitric oxide+ oxygen ➪ nitrogen dioxide

$2NO(g) + O_2(g) \Rightarrow 2NO_2(g)$

▼ Nitric oxide is a colourless gas that is almost the same density as air. If a gas jar of nitric oxide is put together with a gas jar of air, and the separating cover slip pulled away, the nitric oxide will start to mix with the air (because gas molecules are constantly moving) and oxidise to brown nitrogen dioxide. Because nitrogen dioxide is heavier than air, it starts to sink.

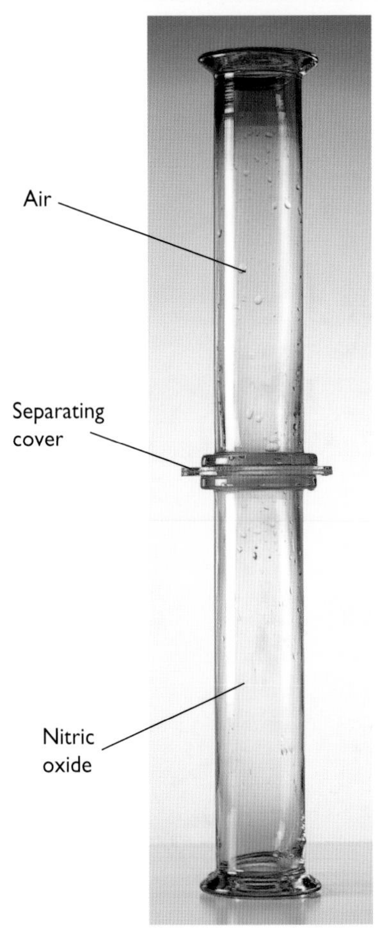

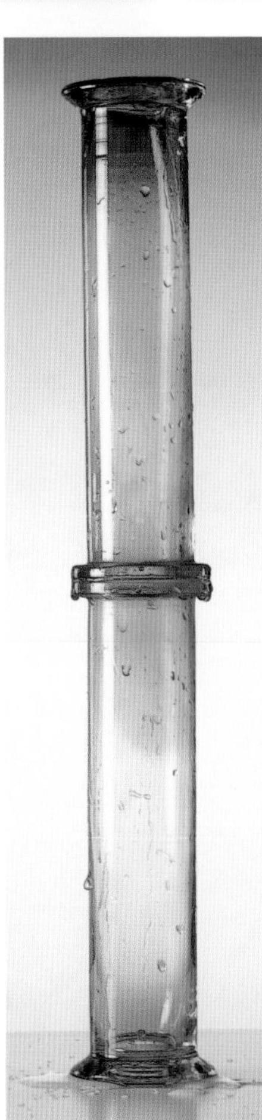

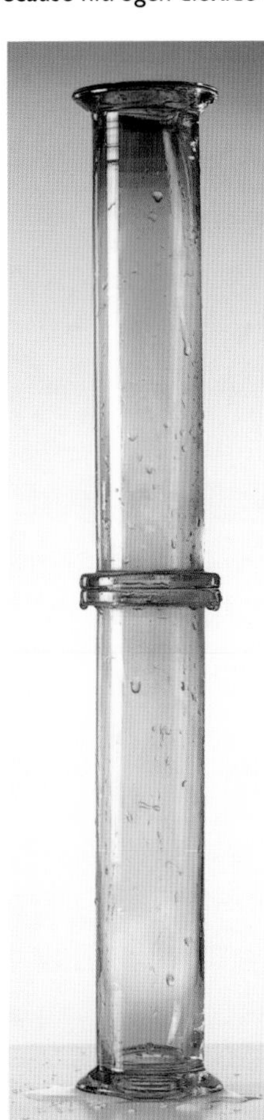

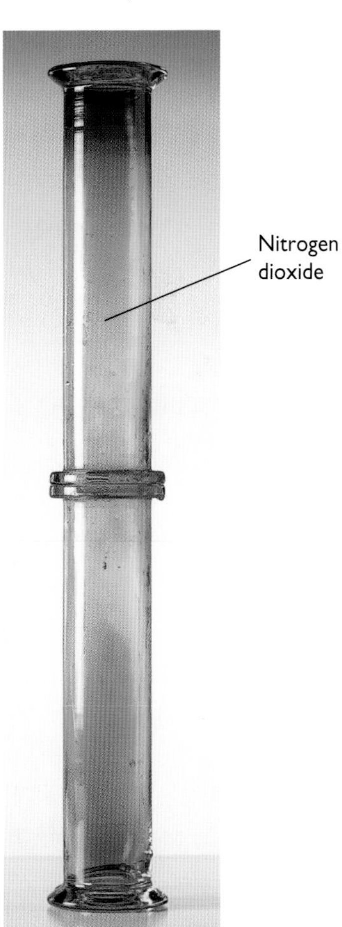

Nitric acid

Nitric acid (once called *aqua fortis*, or "strong water"), a colourless liquid, is the most important nitrogen acid, and has been known about for many centuries. Concentrated nitric acid is an oxidising chemical. The oxygen it contains is readily given up to other reactants.

Nitric acid attacks all common metals except aluminium and iron. By mixing nitric acid and hydrochloric acid, HCl, to give a mixture known as *aqua regia*, it is possible to dissolve all metals, including gold and platinum.

In the atmosphere, hydrogen and nitrogen combine, given the energy of a lightning flash, to produce ammonia. This in turn combines with rain and becomes available to green plants as dilute nitric acid. Dilute nitric acid can be used by plants as a source of nitrogen fertiliser.

Nitric acid is prepared industrially from ammonia, and is used to make a wide variety of compounds, from fertilisers and dyes to explosives.

Preparing nitric acid

This apparatus shows the preparation of fuming nitric acid from concentrated sulphuric acid and potassium nitrate.

Hot fuming nitric acid is incredibly corrosive, so you will notice that all of the laboratory equipment being used is glass (there are no rubber stoppers or tubing for example).

The fuming gas produced when some nitric acid decomposes at this very high temperature is nitrogen dioxide and water vapour formed together with oxygen.

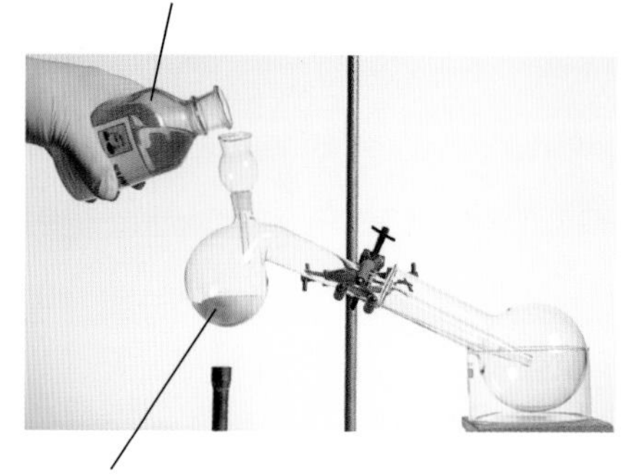

EQUATION: Making fuming nitric acid

Concentrated sulphuric acid + potassium nitrate ⇨ nitric acid + potassium hydrogen sulphate

$H_2SO_4(l)$ + $KNO_3(s)$ ⇨ $HNO_3(g)$ + $KHSO_4(s)$

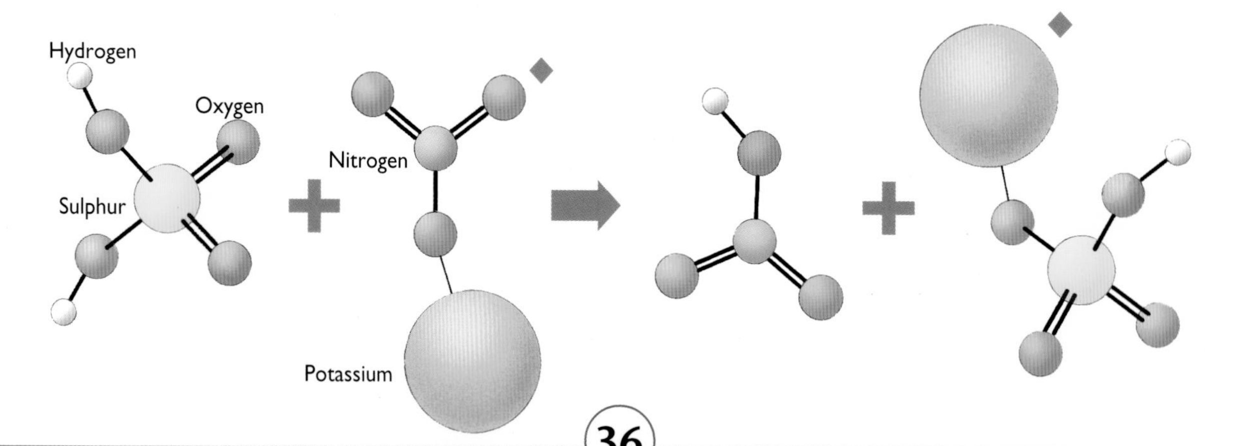

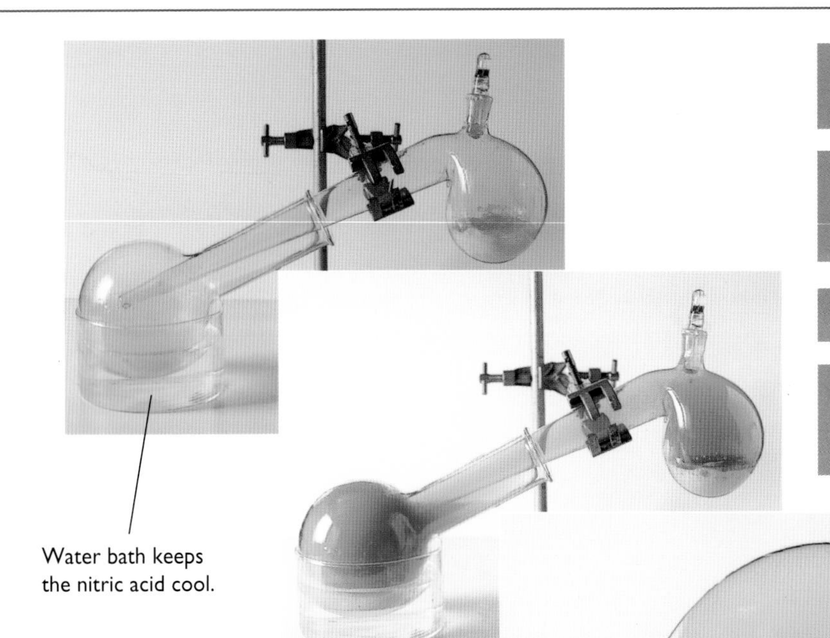

Water bath keeps the nitric acid cool.

fuming: an unstable liquid that gives off a gas. Very concentrated acid solutions are often fuming solutions.

oxidising agent: a substance that removes electrons from another substance (and therefore is itself reduced).

reagent: a starting material for a reaction.

vapour: the gaseous form of a substance that is normally a liquid. For example, water vapour is the gaseous form of liquid water.

Fuming nitric acid

The products of nitric acid

Nitric acid reacts with a wide variety of substances. The reaction oxidises the other reagent, and reduces the nitric acid to a range of products from nitrogen dioxide to ammonia.

Very dilute nitric acid reacts with reactive metals such as magnesium to produce hydrogen and nitric oxide. Less reactive metals such as zinc and iron produce ammonia. Poorly reactive metals such as copper and mercury result in the production of nitric oxide.

Concentrated nitric acid produces nitrogen dioxide instead of nitric oxide. The difference is caused by the fact that concentrated nitric acid is able to oxidize the nitric oxide released, thus forming nitrogen dioxide.

A few metals, such as gold and platinum, are not affected by nitric acid, while others, such as aluminium, do not react because they are protected by their oxide coating.

Also... Making nitric acid industrially

Ammonia (along with oxygen) is used to make nitric acid. However, ammonia and oxygen do not combine easily, and a three stage process has to be employed.

Stage 1: oxygen and ammonia combine to make nitric oxide (NO). This is achieved at high temperatures and uses a catalyst of platinum gauze. The reaction takes place in less than one-thousandth of a second.

Stage 2: Nitric oxide and oxygen are combined to produce nitrogen dioxide (NO_2).

Stage 3: Nitrogen dioxide is absorbed into water to make nitric acid.

The whole process is called the Ostwald process, and it shows that, on some occasions, in practical terms what appears to be a rather simple reaction is in fact quite difficult to achieve.

White phosphorus

Phosphorus is a highly reactive element and does not occur uncombined in nature. However, it is widely found in compounds in rocks, especially in light-coloured rocks called apatites (like calcium phosphate).

Three main forms of phosphorus are known: white, red and black phosphorus.

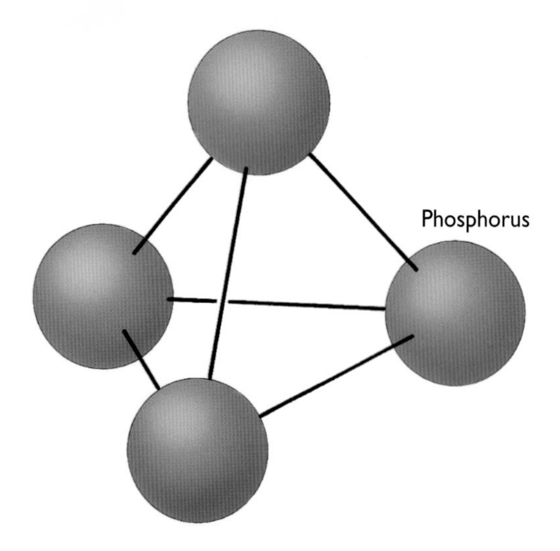

▲▼ White phosphorus. The phosphorus molecule is highly reactive because the bond angle of 60° sets up great strains that can only be reduced by reaction.

White phosphorus

Apatite, sand and carbon react at high temperatures to produce phosphorus and carbon monoxide gases and leave behind a slag of calcium silicate.

Phosphorus gas is collected by condensation and is made into a soft, white, waxy solid called white phosphorus.

One interesting property of white phosphorus is that it gives out a green glow that can easily be seen in the dark. This is caused by the reaction of phosphorus vapours with air.

White phosphorus can catch fire spontaneously in air and has to be stored under water. The main uses of white phosphorus are for incendiary (fire-starting) bombs, napalm bombs and rat poison.

In the past, white phosphorus was also used for strike-anywhere matches, also called "Lucifers". White phosphorus is highly poisonous and many people working in the match-making industry used to suffer from "fossy jaw", until safer forms of phosphorus were introduced (see next page). Modern strike-anywhere matches contain a phosphorus compound mixed with a source of oxygen, potassium chlorate. Striking them against a surface heats the mixture, which then reacts and ignites.

Preparing phosphorus

Phosphorus is prepared by heating crushed calcium phosphate rock with coke and sand. This reduces the calcium phosphate to phosphorus.

The phosphorus vapour is then condensed and made into a liquid.

EQUATION: Preparation of phosphorus from calcium phosophate

Calcium phosphate + carbon + silica ⇨ phosphorus vapour + carbon monoxide gas + calcium silicate (slag)

$Ca_3(PO_4)_2(s) + 5C(s) + 3SiO_2(s) \Rightarrow P_2(g) + 5CO(g) + 3CaSiO_3(s)$

▼▶ Phosphorus reacts with alkalis, which is an unusual property. When white phosphorus is placed in a sodium hydroxide solution, bubbles of gas form. The gas is (poisonous) phosphine gas (PH_3) that has a fishy smell. Phosphine gas is prepared in the absence of air in this apparatus. The phosphine gas contains an impurity, diphosphine (P_2H_4), which combusts spontaneously in air. This creates the flame you see.

The phosphine gas bubbles catch fire as they burst on the surface of the water. The reaction with the oxygen in the air as it burns forms solid phosphorus oxide and creates white "smoke-rings".

combustion: the special case of oxidisation of a substance where a considerable amount of heat and usually light are given out. Combustion is often referred to as "burning".

spontaneous combustion: the effect of a very reactive material beginning to oxidise very quickly and bursting into flame.

vapour: the gaseous form of a substance that is normally a liquid. For example, water vapour is the gaseous form of liquid water.

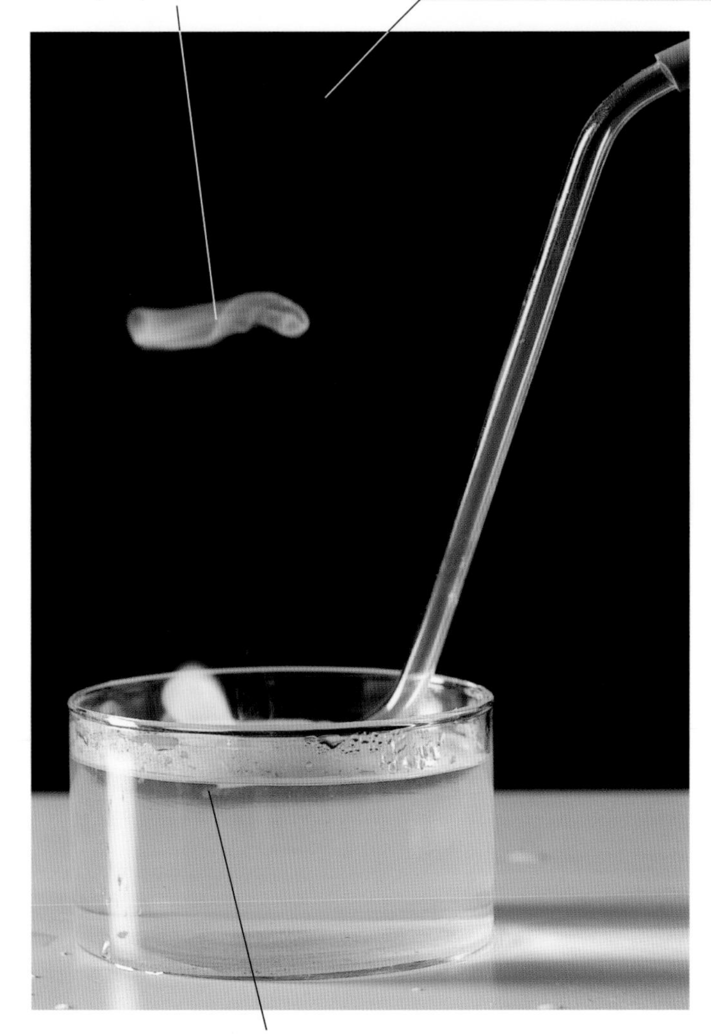

"Smoke-ring" of phosphorus oxide

Bubbles of phosphine and diphosphine gas pass through the water and ignite as they burst on reaching the surface.

Concentrated sodium hydroxide

White phosphorus

Red phosphorus

When white phosphorus is heated in the absence of air it changes to red phosphorus. This is much less reactive than white phosphorus and only bursts into flame when it is heated by friction. The main use of red phosphorus is for safety match heads, tracer bullets, smoke screens and skywriting.

Matches

Matches consist of an incendiary (fire-making) device and a source of fuel, the wood of the match stick. The first matches were tipped with sulphur but these needed a spark from a flint in a tinder box to light them.

Friction matches – those that ignite simply by being rubbed against a rough surface – were invented in the 1830s. The "strike anywhere" matches were tipped with mixtures that included the incendiary chemical white phosphorus that could ignite by spontaneous combustion. These are the matches that will ignite when scraped against any surface. However, many accidents were caused by accidental scraping, so today a more stable (and nonpoisonous) chemical containing phosphorus is used.

Safety matches use red phosphorus (on the side of the match box, not in the match head) for ignition. The match head contains potassium chlorate (a source of oxygen). When the match is struck, the head scrapes across the phosphorus on the match box side, causing a small amount of red phosphorus to heat up and come into contact with a source of oxygen. It then reacts and ignites, in turn causing the match head to begin to burn for long enough to catch the wood of the stick alight.

◀ Red phosphorus.

▶ The wood provides the fuel for the match.

▼▶ These "strike anywhere" matches contain a mixture of tetraphosphorus trisulphide (P_4S_3) and potassium chlorate ($KClO_3$). "Safety" matches use potassium chlorate in the head of the match and red phosphorus in the striker on the match-box side. The potassium chlorate provides the oxygen for the reaction.

combustion: the special case of oxidisation of a substance where a considerable amount of heat and usually light are given out. Combustion is often referred to as "burning".

incendiary: a substance designed to cause burning.

spontaneous combustion: the effect of a very reactive material beginning to oxidise very quickly and bursting into flame.

Phosphoric acid and phosphates

Phosphorus, in the form of compounds called phosphates, occur in all living things. They play a vital role in allowing both plants and animals to transfer energy.

Calcium phosphate is also important to bone-making, comprising about 1.5 kilograms of every adult human skeleton. Phosphates are also found in the chromosomes involved in reproduction.

Because of its crucial role to life, phosphorus is a key element in fertilisers. Bone meal, crushed animal bones, is a traditional way of returning phosphorus to the soil. The alternative supply is from phosphate rock (apatite), which is mined in Florida and Morocco. This is crushed and treated to produce the fertiliser called superphosphate.

Using phosphoric acids

When phosphoric acid is reacted with bases a range of useful products results.

When the base contains sodium (as in sodium hydroxide) the resulting salt is trisodium phosphate, used in washing powders for many years.
When reacted with an organic compound the phosphate is useful as a filler for small cracks.

Phosphoric acid is a hard, glassy solid, the basis of the hard surface for some kinds of enamel.

In very low concentrations, phosphoric acid is also used to produce the tart taste in some soft drinks.

EQUATION: Phosphoric acid rust-inhibitor

Phosphoric acid + iron oxide ⇨ iron phosphate + water

$$2H_3PO_4(aq) + Fe_2O_3(s) \Rightarrow 2FePO_4(s) + 3H_2O(aq)$$

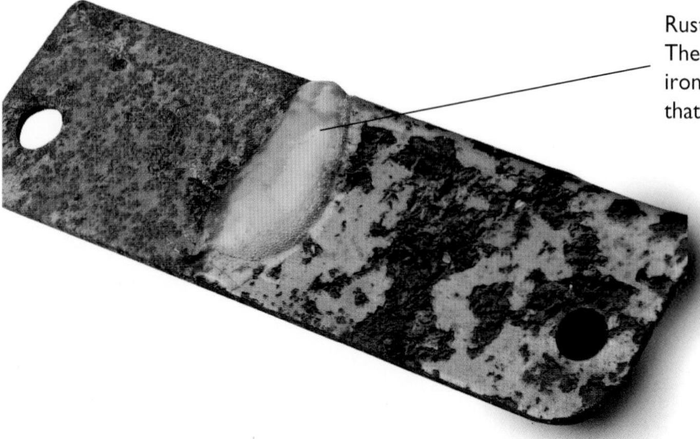

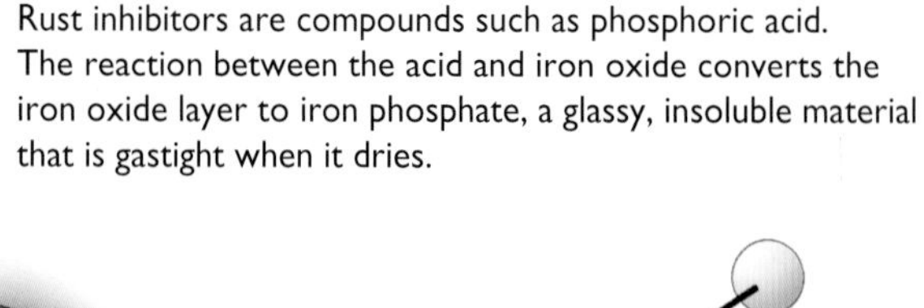

Rust inhibitors are compounds such as phosphoric acid. The reaction between the acid and iron oxide converts the iron oxide layer to iron phosphate, a glassy, insoluble material that is gastight when it dries.

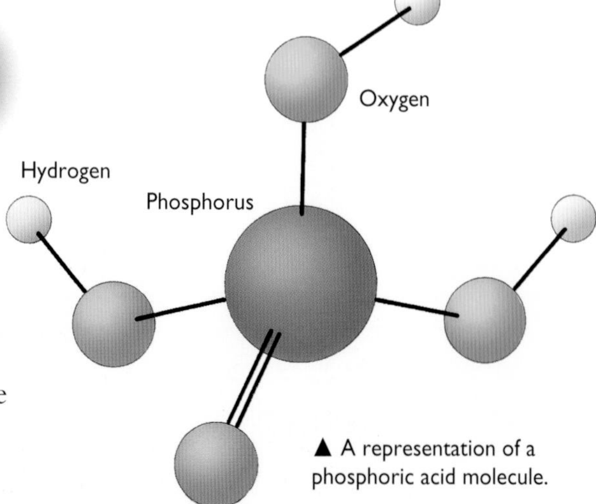

▲ A representation of a phosphoric acid molecule.

Preparing phosphoric acid

Crushed phosphate rock (calcium phosphate) is reacted with sulphuric acid (as in making superphosphate). Further treatment with sulphuric acid, and removal of the gypsum precipitate by filtering, leaves a 60–70% solution of phosphoric acid.

Superphosphate

This is the most important salt derived from phosphoric acid. In the most common form of preparation, phosphate rock is treated with (inexpensive) sulphuric acid. This produces the products superphosphate and gypsum (calcium sulphate), although as the gypsum has no harmful effects it is not removed.

If calcium phosphate is treated with phosphoric acid instead of sulphuric acid, only superphosphate is produced. This more concentrated form of fertiliser is called double or triple phosphate.

detergent: a petroleum-based chemical that removes dirt.

ion: an atom, or group of atoms, that has gained or lost one or more electrons and so developed an electrical charge.

nutrients: soluble ions that are essential to life.

precipitate: tiny solid particles formed as a result of a chemical reaction between two liquids or gases.

EQUATION: Preparation of superphosphate

Calcium phosphate + sulphuric acid ⇨ calcium phosphate + calcium sulphate

$Ca_3(PO_4)_2(s) + 2H_2SO_4(l) \Rightarrow Ca(H_2PO_4)_2(s) + 2Ca(SO)_4(s)$

superphosphate

▼ These tablets contain phosphate fertilisers. Superphosphates are used widely in agriculture, but although they are excellent fertilisers, when phosphate-rich waters seep through soils to rivers, they can cause just the same environmental damage that caused the removal of phosphates from detergents.

▼ Phosphates formed the basis of detergents for many years, but they are now less frequently used because of the environmental damage they can do.

Phosphates and pollution

Phosphates are not stored in the soil, so they have to be applied quite frequently. Not all the fertiliser applied is used by plants; some drains away with rainwater and reaches rivers. Here water life, especially algae, make use of the increased supply of phosphate fertiliser by growing vigorously.

Sodium triphosphate has been used in detergents because the phosphate ion acts to stop dirt particles that have been removed by detergents from sticking back on to clothes. As the phosphates are washed away through the drains, they eventually reach water supplies where they, too, act as nutrients for algae.

The combined effects of extra sources of phosphates in the world's natural drainage basins has caused widespread alarm because it causes excessive algal growth (called algal blooms) that take the oxygen from the water they have been growing in and so make life almost impossible for all other water creatures. For this reason, few detergents now contain phosphates. Farm fertilisers seeping to waterways, however, remain a serious pollution problem.

Key facts about...

Nitrogen

A colourless gas, chemical symbol N

Essential to life

Not very reactive

Has no taste

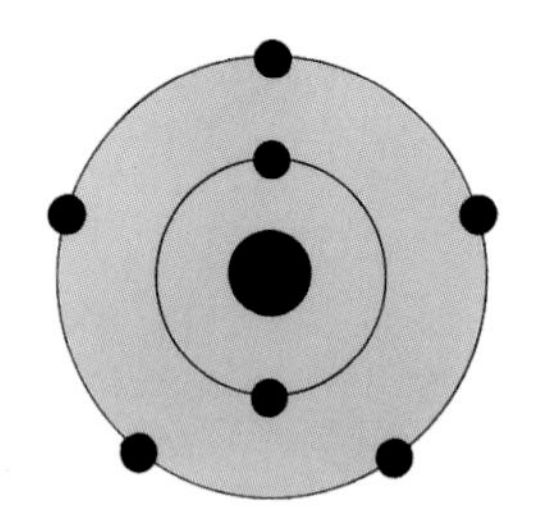

Comprises about 78% of the Earth's atmosphere by volume and 76% by mass (weight)

One of the few elements which is found uncombined in nature

Oxides of nitrogen are major sources of atmospheric pollution

Essential part of many explosives

Atomic number 7, atomic weight about 14

Phosphorus

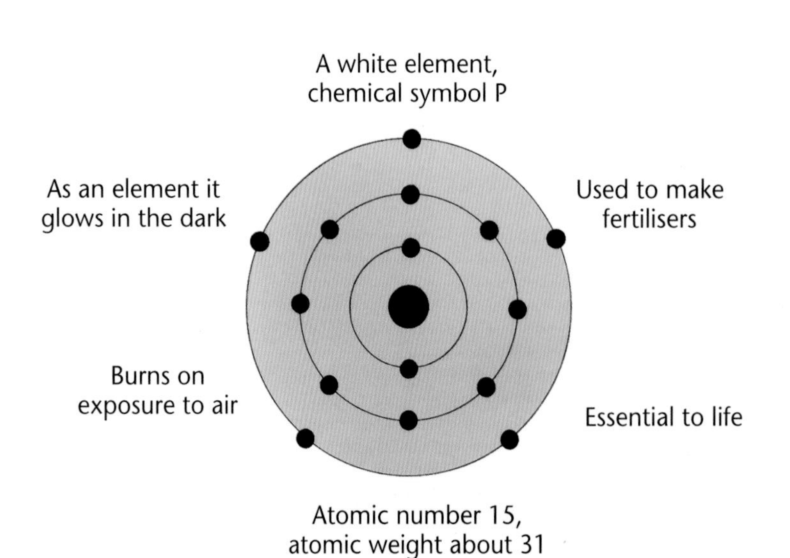

SHELL DIAGRAMS

The shell diagrams on this page are representations of an atom of each element. The total number of electrons are shown in the relevant orbitals, or shells, around the central nucleus.

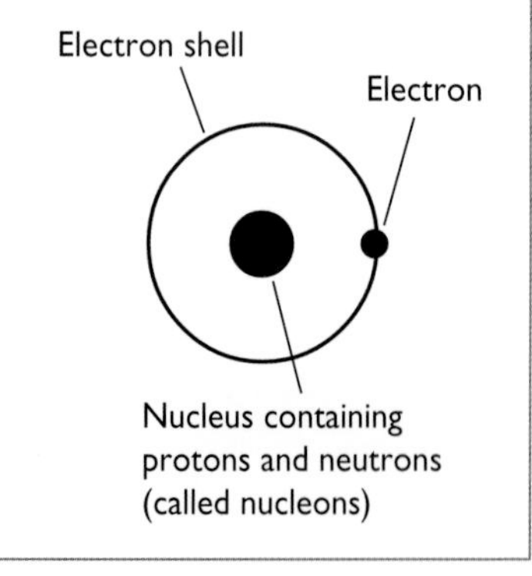

▼ A close up of the inverted flask used in the ammonia fountain experiment shown on page 15.

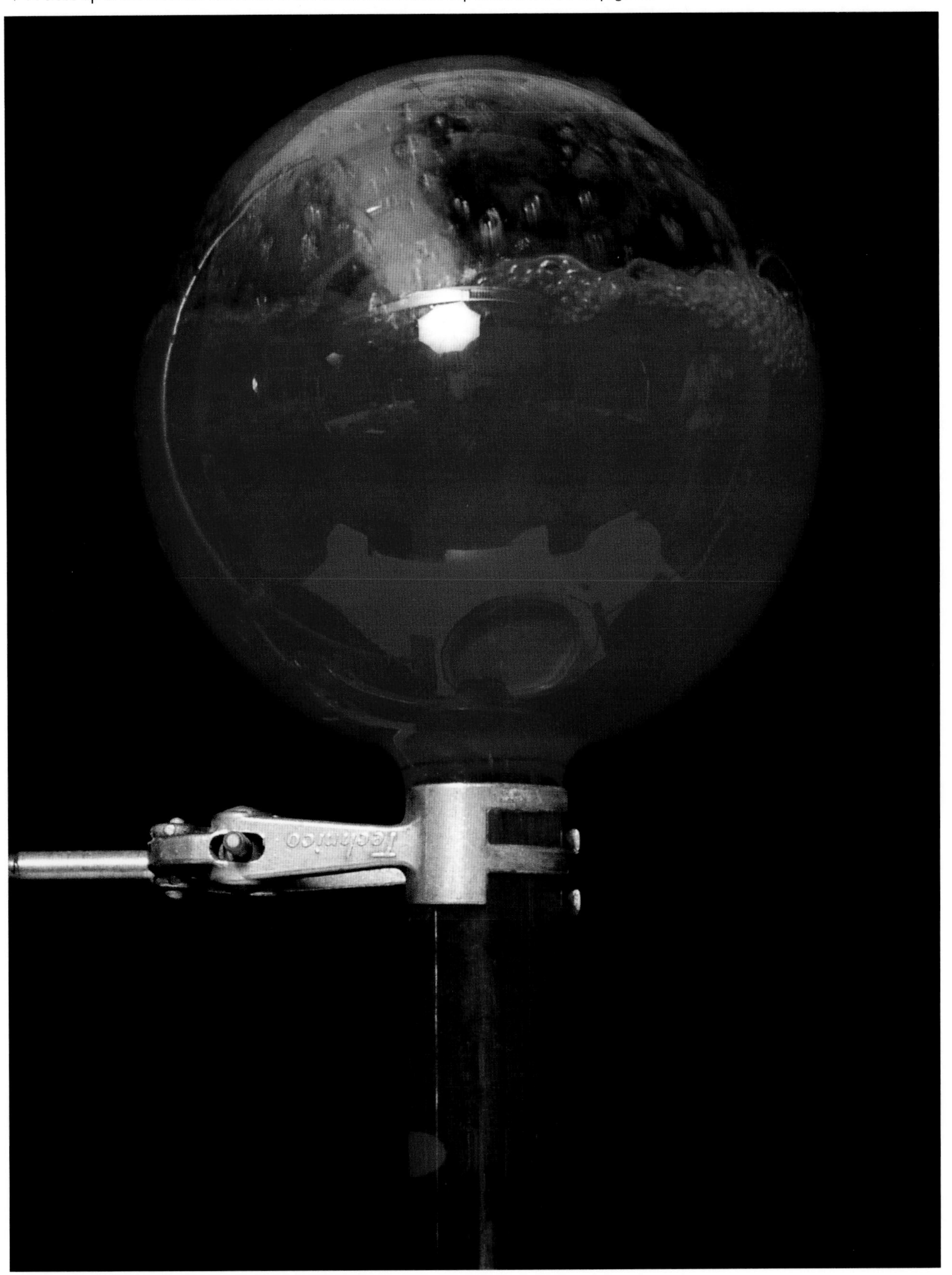

The Periodic Table

The Periodic Table sets out the relationships among the elements of the Universe. According to the Periodic Table, certain elements fall into groups. The pattern of these groups has, in the past, allowed scientists to predict elements that had not at that time been discovered. It can still be used today to predict the properties of unfamiliar elements.

The Periodic Table was first described by a Russian teacher, Dmitry Ivanovich Mendeleev, between 1869 and 1870. He was interested in writing a chemistry textbook, and wanted to show his students that there were certain patterns in the elements that had been discovered. So he set out the elements (of which there were 57 at the time) according to their known properties. On the assumption that there was pattern to the elements, he left blank spaces where elements seemed to be missing. Using this first version of the Periodic Table, he was able to predict in detail the chemical and physical properties of elements that had not yet been discovered. Other scientists began to look for the missing elements, and they soon found them.

GROUP

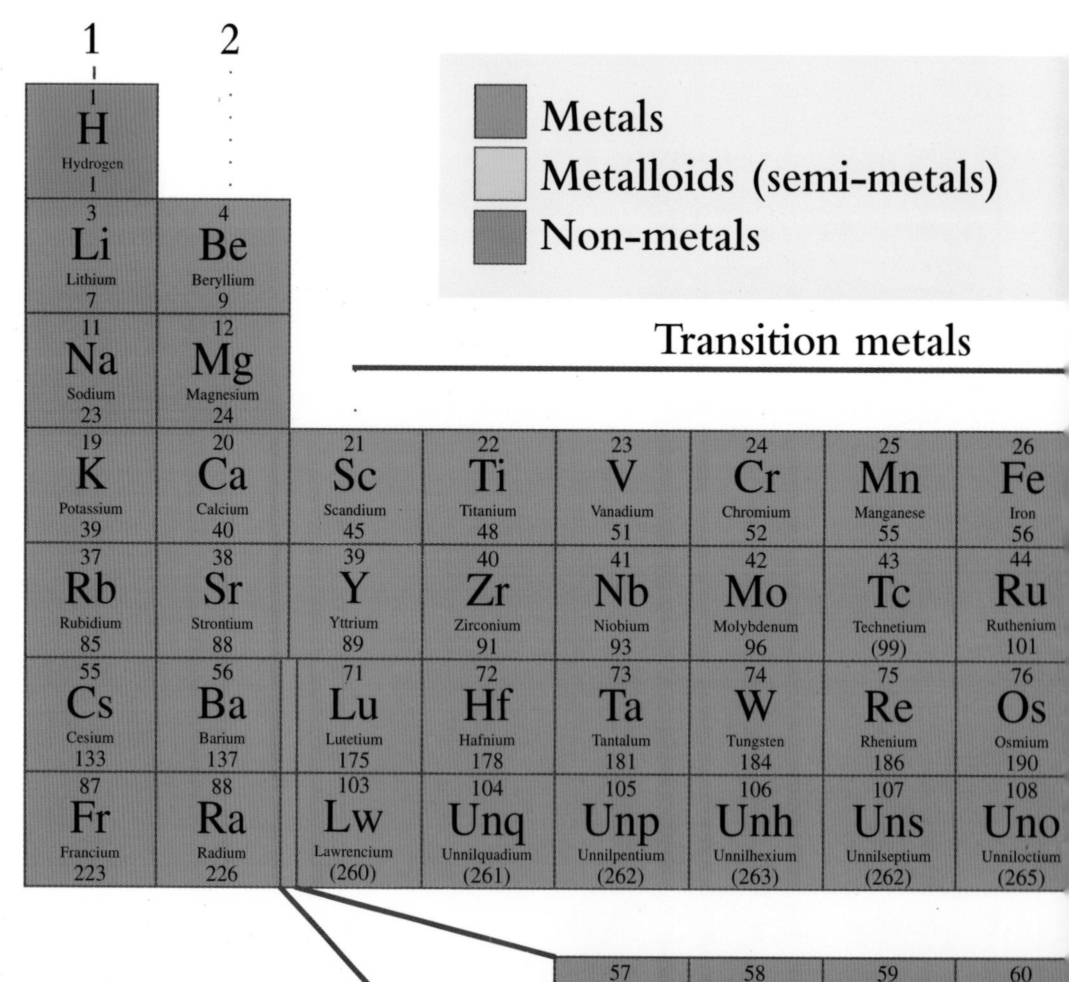

Hydrogen did not seem to fit into the table, so he placed it in a box on its own. Otherwise the elements were all placed horizontally. When an element was reached with properties similar to the first one in the top row, a second row was started. By following this rule, similarities among the elements can be found by reading up and down. By reading across the rows, the elements progressively increase their atomic number. This number indicates the number of positively charged particles (protons) in the nucleus of each atom. This is also the number of negatively charged particles (electrons) in the atom.

The chemical properties of an element depend on the number of electrons in the outermost shell.

Atoms can form compounds by sharing electrons in their outermost shells. This explains why atoms with a full set of electrons (like helium, an inert gas) are unreactive, whereas atoms with an incomplete electron shell (such as chlorine) are very reactive. Elements can also combine by the complete transfer of electrons from metals to non-metals and the compounds formed contain ions.

Radioactive elements lose particles from their nucleus and electrons from their surrounding shells. As a result their atomic number changes and they become new elements.

Atomic (proton) number

13 Al Aluminium 27

Symbol

Name

Approximate relative atomic mass (Approximate atomic weight)

				3	4	5	6	7	0
									2 He Helium 4
				5 B Boron 11	6 C Carbon 12	7 N Nitrogen 14	8 O Oxygen 16	9 F Fluorine 19	10 Ne Neon 20
				13 Al Aluminium 27	14 Si Silicon 28	15 P Phosphorus 31	16 S Sulphur 32	17 Cl Chlorine 35	18 Ar Argon 40
27 Co Cobalt 59	28 Ni Nickel 59	29 Cu Copper 64	30 Zn Zinc 65	31 Ga Gallium 70	32 Ge Germanium 73	33 As Arsenic 75	34 Se Selenium 79	35 Br Bromine 80	36 Kr Krypton 84
45 Rh Rhodium 103	46 Pd Palladium 106	47 Ag Silver 108	48 Cd Cadmium 112	49 In Indium 115	50 Sn Tin 119	51 Sb Antimony 122	52 Te Tellurium 128	53 I Iodine 127	54 Xe Xenon 131
77 Ir Iridium 192	78 Pt Platinum 195	79 Au Gold 197	80 Hg Mercury 201	81 Tl Thallium 204	82 Pb Lead 207	83 Bi Bismuth 209	84 Po Polonium (209)	85 At Astatine (210)	86 Rn Radon (222)
109 Une Unnilennium (266)									

61 Pm Promethium (145)	62 Sm Samarium 150	63 Eu Europium 152	64 Gd Gadolinium 157	65 Tb Terbium 159	66 Dy Dysprosium 163	67 Ho Holmium 165	68 Er Erbium 167	69 Tm Thulium 169	70 Yb Ytterbium 173
93 Np Neptunium (237)	94 Pu Plutonium (244)	95 Am Americium (243)	96 Cm Curium (247)	97 Bk Berkelium (247)	98 Cf Californium (251)	99 Es Einsteinium (252)	100 Fm Fermium (257)	101 Md Mendelevium (258)	102 No Nobelium (259)

Understanding equations

As you read through this book, you will notice that many pages contain equations using symbols. If you are not familiar with these symbols, read this page. Symbols make it easy for chemists to write out the reactions that are occurring in a way that allows a better understanding of the processes involved.

Symbols for the elements

The basis of the modern use of symbols for elements dates back to the 19th century. At this time a shorthand was developed using the first letter of the element wherever possible. Thus "O" stands for oxygen, "H" stands for hydrogen and so on. However, if we were to use only the first letter, then there could be some confusion. For example, nitrogen and nickel would both use the symbols N. To overcome this problem, many elements are symbolised using the first two letters of their full name, and the second letter is lowercase. Thus although nitrogen is N, nickel becomes Ni. Not all symbols come from the English name; many use the Latin name instead. This is why, for example, gold is not G but Au (for the Latin *aurum*) and sodium has the symbol Na, from the Latin *natrium*.

Compounds of elements are made by combining letters. Thus the molecule carbon

Written and symbolic equations

In this book, important chemical equations are briefly stated in words (these are called word equations), and are then shown in their symbolic form along with the states.

What reaction the equation illustrates

Word equation

Symbol equation

EQUATION: The formation of calcium hydroxide

Calcium oxide + water ➪ calcium hydroxide

$CaO(s) + H_2O(l) \Rightarrow Ca(OH)_2(aq)$

heated

Sometimes you will find additional descriptions below the symbolic equation.

Symbol showing the state: *s* is for solid, *l* is for liquid, *g* is for gas and *aq* is for aqueous.

Diagrams

Some of the equations are shown as graphic representations.

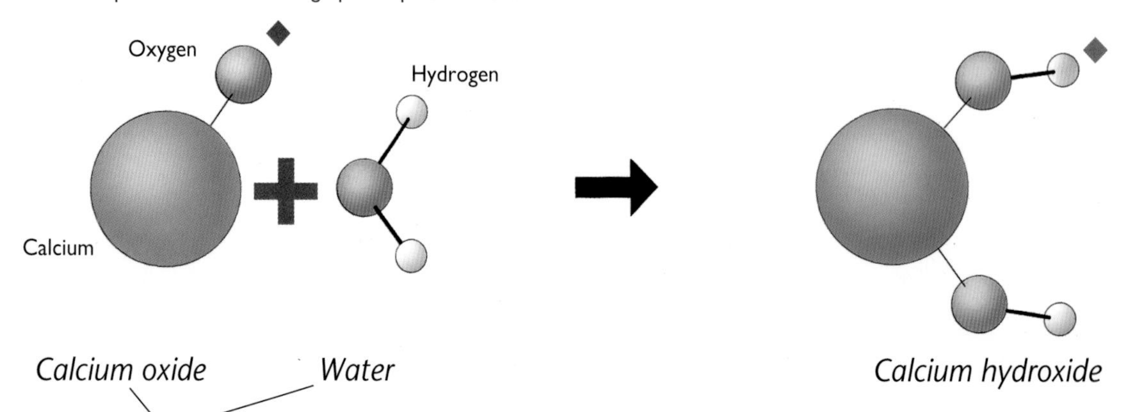

Sometimes the written equation is broken up and put below the relevant stages in the graphic representation.

monoxide is CO. By using lowercase letters for the second letter of an element, it is possible to show that cobalt, symbol Co, is not the same as the molecule carbon monoxide, CO.

However, the letters can be made to do much more than this. In many molecules, atoms combine in unequal numbers. So, for example, carbon dioxide has one atom of carbon for every two of oxygen. This is shown by using the number 2 beside the oxygen, and the symbol becomes CO_2.

In practice, some groups of atoms combine as a unit with other substances. Thus, for example, calcium bicarbonate (one of the compounds used in some antacid pills) is written $Ca(HCO_3)_2$. This shows that the part of the substance inside the brackets reacts as a unit and the "2" outside the brackets shows the presence of two such units.

Some substances attract water molecules to themselves. To show this a dot is used. Thus the blue form of copper sulphate is written $CuSO_4.5H_2O$. In this case five molecules of water attract to one of copper sulphate.

Atoms and ions
Each sphere represents a particle of an element. A particle can be an atom or an ion. Each atom or ion is associated with other atoms or ions through bonds – forces of attraction. The size of the particles and the nature of the bonds can be extremely important in determining the nature of the reaction or the properties of the compound.

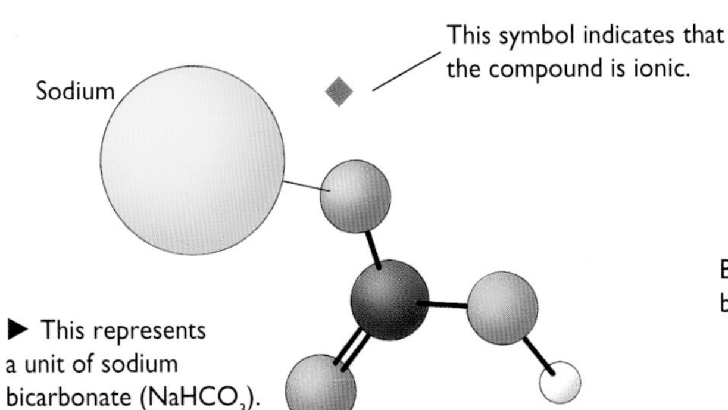

▶ This represents a unit of sodium bicarbonate ($NaHCO_3$).

The term "unit" is sometimes used to simplify the representation of a combination of ions.

When you see the dot, you know that this water can be driven off by heating; it is part of the crystal structure.

In a reaction substances change by rearranging the combinations of atoms. The way they change is shown by using the chemical symbols, placing those that will react (the starting materials, or reactants) on the left and the products of the reaction on the right. Between the two, chemists use an arrow to show which way the reaction is occurring.

It is possible to describe a reaction in words. This gives word equations, which are given throughout this book. However, it is easier to understand what is happening by using an equation containing symbols. These are also given in many places. They are not given when the equations are very complex.

In any equation both sides balance; that is, there must be an equal number of like atoms on both sides of the arrow. When you try to write down reactions, you, too, must balance your equation; you cannot have a few atoms left over at the end!

The symbols in brackets are abbreviations for the physical state of each substance taking part, so that (*s*) is used for solid, (*l*) for liquid, (*g*) for gas and (*aq*) for an aqueous solution, that is, a solution of a substance dissolved in water.

Chemical symbols, equations and diagrams
The arrangement of any molecule or compound can be shown in one of the two ways shown below, depending on which gives the clearer picture. The left-hand diagram is called a ball-and-stick diagram because it uses rods and spheres to show the structure of the material. This example shows water, H_2O. There are two hydrogen atoms and one oxygen atom.

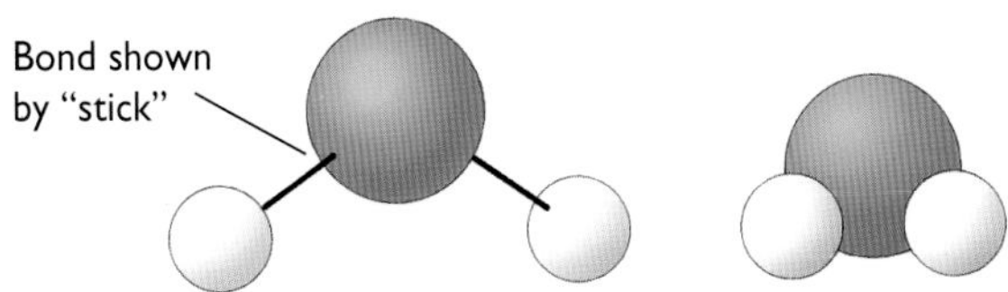

Colours too
The colours of each of the particles help differentiate the elements involved. The diagram can then be matched to the written and symbolic equation given with the diagram. In the case above, oxygen is red and hydrogen is grey.

Glossary of technical terms

absorb: to soak up a substance. Compare to adsorb.

acetone: a petroleum-based solvent.

acid: compounds containing hydrogen which can attack and dissolve many substances. Acids are described as weak or strong, dilute or concentrated, mineral or organic.

acidity: a general term for the strength of an acid in a solution.

acid rain: rain that is contaminated by acid gases such as sulphur dioxide and nitrogen oxides released by pollution.

adsorb/adsorption: to "collect" gas molecules or other particles on to the *surface* of a substance. They are not chemically combined and can be removed. (The process is called "adsorption".) Compare to absorb.

alchemy: the traditional "art" of working with chemicals that prevailed through the Middle Ages. One of the main challenges of alchemy was to make gold from lead. Alchemy faded away as scientific chemistry was developed in the 17th century.

alkali: a base in solution.

alkaline: the opposite of acidic. Alkalis are bases that dissolve, and alkaline materials are called basic materials. Solutions of alkalis have a pH greater than 7.0 because they contain relatively few hydrogen ions.

alloy: a mixture of a metal and various other elements.

alpha particle: a stable combination of two protons and two neutrons, which is ejected from the nucleus of a radioactive atom as it decays. An alpha particle is also the nucleus of the atom of helium. If it captures two electrons it can become a neutral helium atom.

amalgam: a liquid alloy of mercury with another metal.

amino acid: amino acids are organic compounds that are the building blocks for the proteins in the body.

amorphous: a solid in which the atoms are not arranged regularly (i.e. "glassy"). Compare with crystalline.

amphoteric: a metal that will react with both acids and alkalis.

anhydrous: a substance from which water has been removed by heating. Many hydrated salts are crystalline. When they are heated and the water is driven off, the material changes to an anhydrous powder.

anion: a negatively charged atom or group of atoms.

anode: the negative terminal of a battery or the positive electrode of an electrolysis cell.

anodising: a process that uses the effect of electrolysis to make a surface corrosion-resistant.

antacid: a common name for any compound that reacts with stomach acid to neutralise it.

antioxidant: a substance that prevents oxidation of some other substance.

aqueous: a solid dissolved in water. Usually used as "aqueous solution".

atom: the smallest particle of an element.

atomic number: the number of electrons or the number of protons in an atom.

atomised: broken up into a very fine mist. The term is used in connection with sprays and engine fuel systems.

aurora: the "northern lights" and "southern lights" that show as coloured bands of light in the night sky at high latitudes. They are associated with the way cosmic rays interact with oxygen and nitrogen in the air.

basalt: an igneous rock with a low proportion of silica (usually below 55%). It has microscopically small crystals.

base: a compound that may be soapy to the touch and that can react with an acid in water to form a salt and water.

battery: a series of electrochemical cells.

bauxite: an ore of aluminium, of which about half is aluminium oxide.

becquerel: a unit of radiation equal to one nuclear disintegration per second.

beta particle: a form of radiation in which electrons are emitted from an atom as the nucleus breaks down.

bleach: a substance that removes stains from materials either by oxidising or reducing the staining compound.

boiling point: the temperature at which a liquid boils, changing from a liquid to a gas.

bond: chemical bonding is either a transfer or sharing of electrons by two or more atoms. There are a number of types of chemical bond, some very strong (such as covalent bonds), others weak (such as hydrogen bonds). Chemical bonds form because the linked molecule is more stable than the unlinked atoms from which it formed. For example, the hydrogen molecule (H_2) is more stable than single atoms of hydrogen, which is why hydrogen gas is always found as molecules of two hydrogen atoms.

brass: a metal alloy principally of copper and zinc.

brazing: a form of soldering, in which brass is used as the joining metal.

brine: a solution of salt (sodium chloride) in water.

bronze: an alloy principally of copper and tin.

buffer: a chemistry term meaning a mixture of substances in solution that resists a change in the acidity or alkalinity of the solution.

capillary action: the tendency of a liquid to be sucked into small spaces, such as between objects and through narrow-pore tubes. The force to do this comes from surface tension.

catalyst: a substance that speeds up a chemical reaction but itself remains unaltered at the end of the reaction.

cathode: the positive terminal of a battery or the negative electrode of an electrolysis cell.

cathodic protection: the technique of making the object that is to be protected from corrosion into the cathode of a cell. For example, a material, such as steel, is protected by coupling it with a more reactive metal, such as magnesium. Steel forms the cathode and magnesium the anode. Zinc protects steel in the same way.

cation: a positively charged atom or group of atoms.

caustic: a substance that can cause burns if it touches the skin.

cell: a vessel containing two electrodes and an electrolyte that can act as an electrical conductor.

ceramic: a material based on clay minerals, which has been heated so that it has chemically hardened.

chalk: a pure form of calcium carbonate made of the crushed bodies of microscopic sea creatures, such as plankton and algae.

change of state: a change between one of the three states of matter, solid, liquid and gas.

chlorination: adding chlorine to a substance.

cladding: a surface sheet of material designed to protect other materials from corrosion.

clay: a microscopically small plate-like mineral that makes up the bulk of many soils. It has a sticky feel when wet.

combustion: the special case of oxidisation of a substance where a considerable amount of heat and usually light are given out. Combustion is often referred to as "burning".

compound: a chemical consisting of two or more elements chemically bonded together. Calcium atoms can combine with carbon atoms and oxygen atoms to make calcium carbonate, a compound of all three atoms.

condensation nuclei: microscopic particles of dust, salt and other materials suspended in the air, which attract water molecules.

conduction: (i) the exchange of heat (heat conduction) by contact with another object or (ii) allowing the flow of electrons (electrical conduction).

convection: the exchange of heat energy with the surroundings produced by the flow of a fluid due to being heated or cooled.

corrosion: the *slow* decay of a substance resulting from contact with gases and liquids in the environment. The term is often applied to metals. Rust is the corrosion of iron.

corrosive: a substance, either an acid or an alkali, that *rapidly* attacks a wide range of other substances.

cosmic rays: particles that fly through space and bombard all atoms on the Earth's surface. When they interact with the atmosphere they produce showers of secondary particles.

covalent bond: the most common form of strong chemical bonding, which occurs when two atoms *share* electrons.

cracking: breaking down complex molecules into simpler components. It is a term particularly used in oil refining.

crude oil: a chemical mixture of petroleum liquids. Crude oil forms the raw material for an oil refinery.

crystal: a substance that has grown freely so that it can develop external faces. Compare with crystalline, where the atoms are not free to form individual crystals and amorphous where the atoms are arranged irregularly.

crystalline: the organisation of atoms into a rigid "honeycomb-like" pattern without distinct crystal faces.

crystal systems: seven patterns or systems into which all of the world's crystals can be grouped. They are: cubic, hexagonal, rhombohedral, tetragonal, orthorhombic, monoclinic and triclinic.

cubic crystal system: groupings of crystals that look like cubes.

curie: a unit of radiation. The amount of radiation emitted by 1 g of radium each second. (The curie is equal to 37 billion becquerels.)

current: an electric current is produced by a flow of electrons through a conducting solid or ions through a conducting liquid.

decay (radioactive decay): the way that a radioactive element changes into another element decause of loss of mass through radiation. For example uranium decays (changes) to lead.

decompose: to break down a substance (for example by heat or with the aid of a catalyst) into simpler components. In such a chemical reaction only one substance is involved.

dehydration: the removal of water from a substance by heating it, placing it in a dry atmosphere, or through the action of a drying agent.

density: the mass per unit volume (e.g. g/cc).

desertification: a process whereby a soil is allowed to become degraded to a state in which crops can no longer grow, i.e. desert-like. Chemical desertification is usually the result of contamination with halides because of poor irrigation practices.

detergent: a petroleum-based chemical that removes dirt.

diaphragm: a semipermeable membrane – a kind of ultra-fine mesh filter – that will allow only small ions to pass through. It is used in the electrolysis of brine.

diffusion: the slow mixing of one substance with another until the two substances are evenly mixed.

digestive tract: the system of the body that forms the pathway for food and its waste products. It begins at the mouth and includes the stomach and the intestines.

dilute acid: an acid whose concentration has been reduced by a large proportion of water.

diode: a semiconducting device that allows an electric current to flow in only one direction.

disinfectant: a chemical that kills bacteria and other microorganisms.

dissociate: to break apart. In the case of acids it means to break up forming hydrogen ions. This is an example of ionisation. Strong acids dissociate completely. Weak acids are not completely ionised and a solution of a weak acid has a relatively low concentration of hydrogen ions.

dissolve: to break down a substance in a solution without a resultant reaction.

distillation: the process of separating mixtures by condensing the vapours through cooling.

doping: adding metal atoms to a region of silicon to make it semiconducting.

dye: a coloured substance that will stick to another substance, so that both appear coloured.

electrode: a conductor that forms one terminal of a cell.

electrolysis: an electrical–chemical process that uses an electric current to cause the break up of a compound and the movement of metal ions in a solution. The process happens in many natural situations (as for example in rusting) and is also commonly used in industry for purifying (refining) metals or for plating metal objects with a fine, even metal coating.

electrolyte: a solution that conducts electricity.

electron: a tiny, negatively charged particle that is part of an atom. The flow of electrons through a solid material such as a wire produces an electric current.

electroplating: depositing a thin layer of a metal onto the surface of another substance using electrolysis.

element: a substance that cannot be decomposed into simpler substances by chemical means

emulsion: tiny droplets of one substance dispersed in another. A common oil in water emulsion is milk. The tiny droplets in an emulsion tend to come together, so another stabilising substance is often needed to wrap the particles of grease and oil in a stable coat. Soaps and detergents are such agents. Photographic film is an example of a solid emulsion.

endothermic reaction: a reaction that takes heat from the surroundings. The reaction of carbon monoxide with a metal oxide is an example.

enzyme: organic catalysts in the form of proteins in the body that speed up chemical reactions. Every living cell contains hundreds of enzymes, which ensure that the processes of life continue. Should enzymes be made inoperative, such as through mercury poisoning, then death follows.

ester: organic compounds, formed by the reaction of an alcohol with an acid, which often have a fruity taste.

evaporation: the change of state of a liquid to a gas. Evaporation happens below the boiling point and is used as a method of separating out the materials in a solution.

exothermic reaction: a reaction that gives heat to the surroundings. Many oxidation reactions, for example, give out heat.

explosive: a substance which, when a shock is applied to it, decomposes very rapidly, releasing a very large amount of heat and creating a large volume of gases as a shock wave.

extrusion: forming a shape by pushing it through a die. For example, toothpaste is extruded through the cap (die) of the toothpaste tube.

fallout: radioactive particles that reach the ground from radioactive materials in the atmosphere.

fat: semi-solid energy-rich compounds derived from plants or animals and which are made of carbon, hydrogen and oxygen. Scientists call these esters.

feldspar: a mineral consisting of sheets of aluminium silicate. This is the mineral from which the clay in soils is made.

fertile: able to provide the nutrients needed for unrestricted plant growth.

filtration: the separation of a liquid from a solid using a membrane with small holes.

fission: the breakdown of the structure of an atom, popularly called "splitting the atom" because the atom is split into approximately two other nuclei. This is different from, for example, the small change that happens when radioactivity is emitted.

fixation of nitrogen: the processes that natural organisms, such as bacteria, use to turn the nitrogen of the air into ammonium compounds.

fixing: making solid and liquid nitrogen-containing compounds from nitrogen gas. The compounds that are formed can be used as fertilisers.

fluid: able to flow; either a liquid or a gas.

fluorescent: a substance that gives out visible light when struck by invisible waves such as ultraviolet rays.

flux: a material used to make it easier for a liquid to flow. A flux dissolves metal oxides and so prevents a metal from oxidising while being heated.

foam: a substance that is sufficiently gelatinous to be able to contain bubbles of gas. The gas bulks up the substance, making it behave as though it were semi-rigid.

fossil fuels: hydrocarbon compounds that have been formed from buried plant and animal remains. High pressures and temperatures lasting over millions of years are required. The fossil fuels are coal, oil and natural gas.

fraction: a group of similar components of a mixture. In the petroleum industry the light fractions of crude oil are those with the smallest molecules, while the medium and heavy fractions have larger molecules.

free radical: a very reactive atom or group with a "spare" electron.

freezing point: the temperature at which a substance changes from a liquid to a solid. It is the same temperature as the melting point.

fuel: a concentrated form of chemical energy. The main sources of fuels (called fossil fuels because they were formed by geological processes) are coal, crude oil and natural gas. Products include methane, propane and gasoline. The fuel for stars and space vehicles is hydrogen.

fuel rods: rods of uranium or other radioactive material used as a fuel in nuclear power stations.

fuming: an unstable liquid that gives off a gas. Very concentrated acid solutions are often fuming solutions.

fungicide: any chemical that is designed to kill fungi and control the spread of fungal spores.

fusion: combining atoms to form a heavier atom.

galvanising: applying a thin zinc coating to protect another metal.

gamma rays: waves of radiation produced as the nucleus of a radioactive element rearranges itself into a tighter cluster of protons and neutrons. Gamma rays carry enough energy to damage living cells.

gangue: the unwanted material in an ore.

gas: a form of matter in which the molecules form no definite shape and are free to move about to fill any vessel they are put in.

gelatinous: a term meaning made with water. Because a gelatinous precipitate is mostly water, it is of a similar density to water and will float or lie suspended in the liquid.

gelling agent: a semi-solid jelly-like substance.

gemstone: a wide range of minerals valued by people, both as crystals (such as emerald) and as decorative stones (such as agate). There is no single chemical formula for a gemstone.

glass: a transparent silicate without any crystal growth. It has a glassy lustre and breaks with a curved fracture. Note that some minerals have all these features and are therefore natural glasses. Household glass is a synthetic silicate.

glucose: the most common of the natural sugars. It occurs as the polymer known as cellulose, the fibre in plants. Starch is also a form of glucose. The breakdown of glucose provides the energy that animals need for life.

granite: an igneous rock with a high proportion of silica (usually over 65%). It has well-developed large crystals. The largest pink, grey or white crystals are feldspar.

Greenhouse Effect: an increase of the global air temperature as a result of heat released from burning fossil fuels being absorbed by carbon dioxide in the atmosphere.

gypsum: the name for calcium sulphate. It is commonly found as Plaster of Paris and wallboards.

half-life: the time it takes for the radiation coming from a sample of a radioactive element to decrease by half.

halide: a salt of one of the halogens (fluorine, chlorine, bromine and iodine).

halite: the mineral made of sodium chloride.

halogen: one of a group of elements including chlorine, bromine, iodine and fluorine.

heat-producing: see exothermic reaction.

high explosive: a form of explosive that will only work when it receives a shock from another explosive. High explosives are much more powerful than ordinary explosives. Gunpowder is not a high explosive.

hydrate: a solid compound in crystalline form that contains molecular water. Hydrates commonly form when a solution of a soluble salt is evaporated. The water that forms part of a hydrate crystal is known as the "water of crystallization". It can usually be removed by heating, leaving an anhydrous salt.

hydration: the absorption of water by a substance. Hydrated materials are not "wet" but remain firm, apparently dry, solids. In some cases, hydration makes the substance change colour, in many other cases there is no colour change, simply a change in volume.

hydrocarbon: a compound in which only hydrogen and carbon atoms are present. Most fuels are hydrocarbons, as is the simple plastic polyethene (known as polythene).

hydrogen bond: a type of attractive force that holds one molecule to another. It is one of the weaker forms of intermolecular attractive force.

hydrothermal: a process in which hot water is involved. It is usually used in the context of rock formation because hot water and other fluids sent outwards from liquid magmas are important carriers of metals and the minerals that form gemstones.

igneous rock: a rock that has solidified from molten rock, either volcanic lava on the Earth's surface or magma deep underground. In either case the rock develops a network of interlocking crystals.

incendiary: a substance designed to cause burning.

indicator: a substance or mixture of substances that change colour with acidity or alkalinity.

inert: nonreactive.

infra-red radiation: a form of light radiation where the wavelength of the waves is slightly longer than visible light. Most heat radiation is in the infra-red band.

insoluble: a substance that will not dissolve.

ion: an atom, or group of atoms, that has gained or lost one or more electrons and so developed an electrical charge. Ions behave differently from electrically neutral atoms and molecules. They can move in an electric field, and they can also bind strongly to solvent molecules such as water. Positively charged ions are called cations; negatively charged ions are called anions. Ions carry electrical current through solutions.

ionic bond: the form of bonding that occurs between two ions when the ions have opposite charges. Sodium cations bond with chloride anions to form common salt (NaCl) when a salty solution is evaporated. Ionic bonds are strong bonds except in the presence of a solvent.

ionise: to break up neutral molecules into oppositely charged ions or to convert atoms into ions by the loss of electrons.

ionisation: a process that creates ions.

irrigation: the application of water to fields to help plants grow during times when natural rainfall is sparse.

isotope: atoms that have the same number of protons in their nucleus, but which have different masses; for example, carbon-12 and carbon-14.

latent heat: the amount of heat that is absorbed or released during the process of changing state between gas, liquid or solid. For example, heat is absorbed when a substance melts and it is released again when the substance solidifies.

latex: (the Latin word for "liquid") a suspension of small polymer particles in water. The rubber that flows from a rubber tree is a natural latex. Some synthetic polymers are made as latexes, allowing polymerisation to take place in water.

lava: the material that flows from a volcano.

limestone: a form of calcium carbonate rock that is often formed of lime mud. Most limestones are light grey and have abundant fossils.

liquid: a form of matter that has a fixed volume but no fixed shape.

lode: a deposit in which a number of veins of a metal found close together.

lustre: the shininess of a substance.

magma: the molten rock that forms a balloon-shaped chamber in the rock below a volcano. It is fed by rock moving upwards from below the crust.

marble: a form of limestone that has been "baked" while deep inside mountains. This has caused the limestone to melt and reform into small interlocking crystals, making marble harder than limestone.

mass: the amount of matter in an object. In everyday use, the word weight is often used to mean mass.

melting point: the temperature at which a substance changes state from a solid to a liquid. It is the same as freezing point.

membrane: a thin flexible sheet. A semipermeable membrane has microscopic holes of a size that will selectively allow some ions and molecules to pass through but hold others back. It thus acts as a kind of sieve.

meniscus: the curved surface of a liquid that forms when it rises in a small bore, or capillary tube. The meniscus is convex (bulges upwards) for mercury and is concave (sags downwards) for water.

metal: a substance with a lustre, the ability to conduct heat and electricity and which is not brittle.

metallic bonding: a kind of bonding in which atoms reside in a "sea" of mobile electrons. This type of bonding allows metals to be good conductors and means that they are not brittle

metamorphic rock: formed either from igneous or sedimentary rocks, by heat and or pressure. Metamorphic rocks form deep inside mountains during periods of mountain building. They result from the remelting of rocks during which process crystals are able to grow. Metamorphic rocks often show signs of banding and partial melting.

micronutrient: an element that the body requires in small amounts. Another term is trace element.

mineral: a solid substance made of just one element or chemical compound. Calcite is a mineral because it consists only of calcium carbonate, halite is a mineral because it contains only sodium chloride, quartz is a mineral because it consists of only silicon dioxide.

mineral acid: an acid that does not contain carbon and that attacks minerals. Hydrochloric, sulphuric and nitric acids are the main mineral acids.

mineral-laden: a solution close to saturation.

mixture: a material that can be separated out into two or more substances using physical means.

molecule: a group of two or more atoms held together by chemical bonds.

monoclinic system: a grouping of crystals that look like double-ended chisel blades.

monomer: a building block of a larger chain molecule ("mono" means one, "mer" means part).

mordant: any chemical that allows dyes to stick to other substances.

native metal: a pure form of a metal, not combined as a compound. Native metal is more common in poorly reactive elements than in those that are very reactive.

neutralisation: the reaction of acids and bases to produce a salt and water. The reaction causes hydrogen from the acid and hydroxide from the base to be changed to water. For example, hydrochloric acid reacts with sodium hydroxide to form common salt and water. The term is more generally used for any reaction where the pH changes towards 7.0, which is the pH of a neutral solution.

neutron: a particle inside the nucleus of an atom that is neutral and has no charge.

noncombustible: a substance that will not burn.

noble metal: silver, gold, platinum, and mercury. These are the least reactive metals.

nuclear energy: the heat energy produced as part of the changes that take place in the core, or nucleus, of an element's atoms.

nuclear reactions: reactions that occur in the core, or nucleus of an atom.

nutrients: soluble ions that are essential to life.

octane: one of the substances contained in fuel.

ore: a rock containing enough of a useful substance to make mining it worthwhile.

organic acid: an acid containing carbon and hydrogen.

organic substance: a substance that contains carbon.

osmosis: a process where molecules of a liquid solvent move through a membrane (filter) from a region of low concentration to a region of high concentration of solute.

oxidation: a reaction in which the oxidising agent removes electrons. (Note that oxidising agents do not have to contain oxygen.)

oxide: a compound that includes oxygen and one other element.

oxidise: the process of gaining oxygen. This can be part of a controlled chemical reaction, or it can be the result of exposing a substance to the air, where oxidation (a form of corrosion) will occur slowly, perhaps over months or years.

oxidising agent: a substance that removes electrons from another substance (and therefore is itself reduced).

ozone: a form of oxygen whose molecules contain three atoms of oxygen. Ozone is regarded as a beneficial gas when high in the atmosphere because it blocks ultraviolet rays. It is a harmful gas when breathed in, so low level ozone, which is produced as part of city smog, is regarded as a form of pollution. The ozone layer is the uppermost part of the stratosphere.

pan: the name given to a shallow pond of liquid. Pans are mainly used for separating solutions by evaporation.

patina: a surface coating that develops on metals and protects them from further corrosion.

percolate: to move slowly through the pores of a rock.

period: a row in the Periodic Table.

Periodic Table: a chart organising elements by atomic number and chemical properties into groups and periods.

pesticide: any chemical that is designed to control pests (unwanted organisms) that are harmful to plants or animals.

petroleum: a natural mixture of a range of gases, liquids and solids derived from the decomposed remains of plants and animals.

pH: a measure of the hydrogen ion concentration in a liquid. Neutral is pH 7.0; numbers greater than this are alkaline, smaller numbers are acidic.

phosphor: any material that glows when energized by ultraviolet or electron beams such as in fluorescent tubes and cathode ray tubes. Phosphors, such as phosphorus, emit light after the source of excitation is cut off. This is why they glow in the dark. By contrast, fluorescors, such as fluorite, emit light only while they are being excited by ultraviolet light or an electron beam.

photon: a parcel of light energy.

photosynthesis: the process by which plants use the energy of the Sun to make the compounds they need for life. In photosynthesis, six molecules of carbon dioxide from the air combine with six molecules of water, forming one molecule of glucose (sugar) and releasing six molecules of oxygen back into the atmosphere.

pigment: any solid material used to give a liquid a colour.

placer deposit: a kind of ore body made of a sediment that contains fragments of gold ore eroded from a mother lode and transported by rivers and/or ocean currents.

plastic (material): a carbon-based material consisting of long chains (polymers) of simple molecules. The word plastic is commonly restricted to synthetic polymers.

plastic (property): a material is plastic if it can be made to change shape easily. Plastic materials will remain in the new shape. (Compare with elastic, a property where a material goes back to its original shape.)

plating: adding a thin coat of one material to another to make it resistant to corrosion.

playa: a dried-up lake bed that is covered with salt deposits. From the Spanish word for beach.

poison gas: a form of gas that is used intentionally to produce widespread injury and death. (Many gases are poisonous, which is why many chemical reactions are performed in laboratory fume chambers, but they are a byproduct of a reaction and not intended to cause harm.)

polymer: a compound that is made of long chains by combining molecules (called monomers) as repeating units. ("Poly" means many, "mer" means part).

polymerisation: a chemical reaction in which large numbers of similar molecules arrange themselves into large molecules, usually long chains. This process usually happens when there is a suitable catalyst present. For example, ethene reacts to form polythene in the presence of certain catalysts.

porous: a material containing many small holes or cracks. Quite often the pores are connected, and liquids, such as water or oil, can move through them.

precious metal: silver, gold, platinum, iridium, and palladium. Each is prized for its rarity. This category is the equivalent of precious stones, or gemstones, for minerals.

precipitate: tiny solid particles formed as a result of a chemical reaction between two liquids or gases.

preservative: a substance that prevents the natural organic decay processes from occurring. Many substances can be used safely for this purpose, including sulphites and nitrogen gas.

product: a substance produced by a chemical reaction.

protein: molecules that help to build tissue and bone and therefore make new body cells. Proteins contain amino acids.

proton: a positively charged particle in the nucleus of an atom that balances out the charge of the surrounding electrons

pyrite: "mineral of fire". This name comes from the fact that pyrite (iron sulphide) will give off sparks if struck with a stone.

pyrometallurgy: refining a metal from its ore using heat. A blast furnace or smelter is the main equipment used.

radiation: the exchange of energy with the surroundings through the transmission of waves or particles of energy. Radiation is a form of energy transfer that can happen through space; no intervening medium is required (as would be the case for conduction and convection).

radioactive: a material that emits radiation or particles from the nucleus of its atoms.

radioactive decay: a change in a radioactive element due to loss of mass through radiation. For example uranium decays (changes) to lead.

radioisotope: a shortened version of the phrase radioactive isotope.

radiotracer: a radioactive isotope that is added to a stable, nonradioactive material in order to trace how it moves and its concentration.

reaction: the recombination of two substances using parts of each substance to produce new substances.

reactivity: the tendency of a substance to react with other substances. The term is most widely used in comparing the reactivity of metals. Metals are arranged in a reactivity series.

reagent: a starting material for a reaction.

recycling: the reuse of a material to save the time and energy required to extract new material from the Earth and to conserve non-renewable resources.

redox reaction: a reaction that involves reduction and oxidation.

reducing agent: a substance that gives electrons to another substance. Carbon monoxide is a reducing agent when passed over copper oxide, turning it to copper and producing carbon dioxide gas. Similarly, iron oxide is reduced to iron in a blast furnace. Sulphur dioxide is a reducing agent, used for bleaching bread.

reduction: the removal of oxygen from a substance. See also: oxidation.

refining: separating a mixture into the simpler substances of which it is made. In the case of a rock, it means the extraction of the metal that is mixed up in the rock. In the case of oil it means separating out the fractions of which it is made.

refractive index: the property of a transparent material that controls the angle at which total internal reflection will occur. The greater the refractive index, the more reflective the material will be.

resin: natural or synthetic polymers that can be moulded into solid objects or spun into thread.

rust: the corrosion of iron and steel.

saline: a solution in which most of the dissolved matter is sodium chloride (common salt).

salinisation: the concentration of salts, especially sodium chloride, in the upper layers of a soil due to poor methods of irrigation.

salts: compounds, often involving a metal, that are the reaction products of acids and bases. (Note "salt" is also the common word for sodium chloride, common salt or table salt.)

saponification: the term for a reaction between a fat and a base that produces a soap.

saturated: a state where a liquid can hold no more of a substance. If any more of the substance is added, it will not dissolve.

saturated solution: a solution that holds the maximum possible amount of dissolved material. The amount of material in solution varies with the temperature; cold solutions can hold less dissolved solid material than hot solutions. Gases are more soluble in cold liquids than hot liquids.

sediment: material that settles out at the bottom of a liquid when it is still.

semiconductor: a material of intermediate conductivity. Semiconductor devices often use silicon when they are made as part of diodes, transistors or integrated circuits.

semipermeable membrane: a thin (membrane) of material that acts as a fine sieve, allowing small molecules to pass, but holding large molecules back.

silicate: a compound containing silicon and oxygen (known as silica).

sintering: a process that happens at moderately high temperatures in some compounds. Grains begin to fuse together even through they do not melt. The most widespread example of sintering happens during the firing of clays to make ceramics.

slag: a mixture of substances that are waste products of a furnace. Most slags are composed mainly of silicates.

smelting: roasting a substance in order to extract the metal contained in it.

smog: a mixture of smoke and fog. The term is used to describe city fogs in which there is a large proportion of particulate matter (tiny pieces of carbon from exhausts) and also a high concentration of sulphur and nitrogen gases and probably ozone.

soldering: joining together two pieces of metal using solder, an alloy with a low melting point.

solid: a form of matter where a substance has a definite shape.

soluble: a substance that will readily dissolve in a solvent.

solute: the substance that dissolves in a solution (e.g. sodium chloride in salt water).

solution: a mixture of a liquid and at least one other substance (e.g. salt water). Mixtures can be separated out by physical means, for example by evaporation and cooling.

solvent: the main substance in a solution (e.g. water in salt water).

spontaneous combustion: the effect of a very reactive material beginning to oxidise very quickly and bursting into flame.

stable: able to exist without changing into another substance.

stratosphere: the part of the Earth's atmosphere that lies immediately above the region in which clouds form. It occurs between 12 and 50 km above the Earth's surface.

strong acid: an acid that has completely dissociated (ionised) in water. Mineral acids are strong acids.

sublimation: the change of a substance from solid to gas, or vica versa, without going through a liquid phase.

substance: a type of material, including mixtures.

sulphate: a compound that includes sulphur and oxygen, for example, calcium sulphate or gypsum.

sulphide: a sulphur compound that contains no oxygen.

sulphite: a sulphur compound that contains less oxygen than a sulphate.

surface tension: the force that operates on the surface of a liquid, which makes it act as though it were covered with an invisible elastic film.

suspension: tiny particles suspended in a liquid.

synthetic: does not occur naturally, but has to be manufactured.

tarnish: a coating that develops as a result of the reaction between a metal and substances in the air. The most common form of tarnishing is a very thin transparent oxide coating.

thermonuclear reactions: reactions that occur within atoms due to fusion, releasing an immensely concentrated amount of energy.

thermoplastic: a plastic that will soften, can repeatedly be moulded it into shape on heating and will set into the moulded shape as it cools.

thermoset: a plastic that will set into a moulded shape as it cools, but which cannot be made soft by reheating.

titration: a process of dripping one liquid into another in order to find out the amount needed to cause a neutral solution. An indicator is used to signal change.

toxic: poisonous enough to cause death.

translucent: almost transparent.

transmutation: the change of one element into another.

vapour: the gaseous form of a substance that is normally a liquid. For example, water vapour is the gaseous form of liquid water.

vein: a mineral deposit different from, and usually cutting across, the surrounding rocks. Most mineral and metal-bearing veins are deposits filling fractures. The veins were filled by hot, mineral-rich waters rising upwards from liquid volcanic magma. They are important sources of many metals, such as silver and gold, and also minerals such as gemstones. Veins are usually narrow, and were best suited to hand-mining. They are less exploited in the modern machine age.

viscous: slow moving, syrupy. A liquid that has a low viscosity is said to be mobile.

vitreous: glass-like.

volatile: readily forms a gas.

vulcanisation: forming cross-links between polymer chains to increase the strength of the whole polymer. Rubbers are vulcanised using sulphur when making tyres and other strong materials.

weak acid: an acid that has only partly dissociated (ionised) in water. Most organic acids are weak acids.

weather: a term used by Earth scientists and derived from "weathering", meaning to react with water and gases of the environment.

weathering: the slow natural processes that break down rocks and reduce them to small fragments either by mechanical or chemical means.

welding: fusing two pieces of metal together using heat.

X-rays: a form of very short wave radiation.

Index